talk is cheap

switching to
internet telephones

James E. Gaskin

Beijing • Cambridge • Farnham • Köln • Paris • Sebastopol • Taipei • Tokyo

PREFACE

Skype uses your computer for talking instead of a traditional-style phone, but leads the world in registered broadband phone users: 21 million and counting. They did this by offering free (as in absolutely no money required for their software or service) calls to all other Skype users in the world. Read all about them in Chapter 6.

Choosing between the Vonage-type service and the Skype-type service makes a huge difference in your telephone happiness. Yet most people will benefit from a combination of the two services. This book reveals how to make your broadband phone service(s) work they way you want them to work.

How This Book Is Organized

I ordered the chapters in a logical way that answers the most questions for the most readers. However, I realize that few people (including me) read a technical book cover to cover.

This is your book, so read it how you want to read it. Jump to the spots that interest you or read the chapters in order and get into the flow. Either way you should learn plenty about the new types of telephone options available.

Chapter 1, *How Internet Telephone Calls Work*, provides background information highlighting the differences and similarities between the traditional telephone service available for over 100 years and the new wave of intelligent, inexpensive, and feature-rich broadband phone services.

Chapter 2, *Your Internet Phone*, describes the differences between the two types of broadband phone services.

Chapter 3, *Free Internet Phone Features that You're Paying for Now*, details exactly what you are paying for today with a traditional telephone line, and what your options are if you switch to a broadband phone service provider.

Chapter 4, *Choosing your Internet Phone Equipment*, lists all the equipment you may add to improve your broadband phone service if you wish. Equipment options abound to satisfy every equipment need and budget range.

Chapter 5, *Vonage and Other Broadband Phone Carriers*, drills down deeply into the service offerings from Vonage, the leading broadband phone company following the phone-centric model. Multiple Vonage competitors join the fun as well.

Chapter 6, *Skype and Other Computer-centric Services*, shows the other side of the broadband phone world, including the world's most popular Internet telephone service from Skype.

Chapter 7, *911, Alarms, and Other Outgoing Calls*, discusses how to call family and friends who are still using traditional telephone services. The 911 issue takes center stage so you can see how 911 calls are handled now and how much better emergency calls will be in the future with broadband phone services.

Chapter 8, *Tricks, Tips, and Techniques for Advanced Users*, covers multiple methods to handle extension phones with a new service, improve call quality, and manage your new broadband phone service.

Chapter 9, *Go Wireless*, examines the growing integration of cell and broadband phones, along with new phones for the computer-centric services.

In addition to the physical pages you hold now, extra material awaits you on the Web. First stop? Go to *www.gaskin.com/talk/* for updates on broadband phone company services.

Who This Book Is For

If you're tired of paying far more for a traditional telephone line than you think fair, this book is for you.

If you're curious about "those broadband phones" or wonder how Internet Telephony can help you, this book will answer your questions.

If you're curious about products you've heard about that let you talk to anyone in the world for free, this book shows you how.

If you're curious about new developments in voice communications and wonder when interesting features will appear, this book will tell you.

This book is for people who look at their old-fashioned telephone and want to save money, get more features, or both. I'll show you how to do all that and more.

Conventions Used in This Book

The following typographical conventions are used in this book:

Plain text
: Indicates menu titles, menu options, menu buttons, and keyboard accelerators (such as Alt and Ctrl).

Italic
: Indicates new terms, URLs, email addresses, filenames, file extensions, pathnames, directories, and Unix utilities.

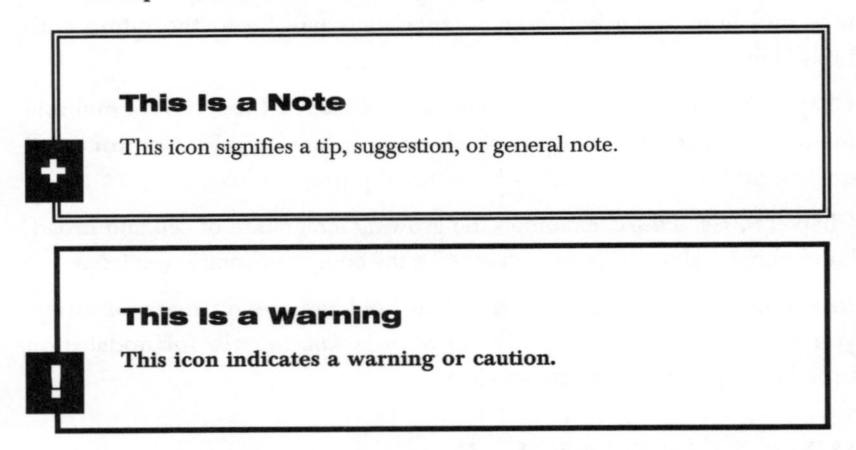

This Is a Note

This icon signifies a tip, suggestion, or general note.

This Is a Warning

This icon indicates a warning or caution.

I'd Like to Hear from You

Send me questions, comments, happy stories, and even tales of woe. I'll read and reply to all. I may not be able to fix your tale of woe, but I will listen. Send email about this book to me at *talk@gaskin.com.*

My main web site is *www.gaskin.com*, and the section devoted to this book is *www.gaskin.com/talk/*. You can read many helpful hints written for home office and small business users at *www.GaskinGuides.com.*

Comments and Questions

Please address comments and questions concerning this book to the publisher:

O'Reilly Media, Inc.
1005 Gravenstein Highway North
Sebastopol, CA 95472
(800) 998-9938 (in the United States or Canada)
(707) 829-0515 (international or local)
(707) 829-0104 (fax)

We have a web page for this book, where we list errata, examples, and any additional information. You can access this page at:

www.oreilly.com/catalog/talk/

To comment or ask technical questions about this book, send email to:

bookquestions@oreilly.com

For more information about our books, conferences, Resource Centers, and the O'Reilly Network, see our web site at:

www.oreilly.com

Safari Enabled

When you see a Safari® Enabled icon on the cover of your favorite technology book, it means the book is available online through the O'Reilly Network Safari Bookshelf.

Safari offers a solution that's better than e-books. It's a virtual library that lets you easily search thousands of top tech books, cut and paste code samples, download chapters, and find quick answers when you need the most accurate, current information. Try it for free at *http://safari.oreilly.com.*

Acknowledgments

There may be a single name on the cover for the author, but every book requires a large and dedicated team. This book is no exception.

Laura Lewin of StudioB put me together with O'Reilly to start this process. Brian Jepson of O'Reilly worked with me to develop, fine-tune, and polish the manuscript. An entire team of O'Reilly production people remain hidden, their anonymity not reducing the amount of hard work required to convert this book from files created with Microsoft Office in my house to pounds of paper in bookstores.

Katherine Foster of Connors Communications enlisted the top executives of Vonage to give me their time for book details. Chief Executive Office Jeffrey Citron, Chief Technical Officer Louie Mamakos, and Executive Vice President of Operations Michael Tribolet all spent more time than necessary speaking with me about the industry in general and Vonage in particular. Kat James of Skype gathered information from all over Europe to ensure I understood how and why Skype chose their method of operation. Jupiter Research analyst Joseph Laszlo provided industry perspective and Julia Di Dominicus, Director of Business Development at The Hub, one of New York's most advanced data- and telecom-networking facilities, gave me valuable industry insider information.

Thanks to my buddy Ed Tittel for early positive reinforcement on the contents and tone of this book. And thanks to Philip LeNir for jumping in to provide a good technical backstop and detailed Skype information.

Dedication

As always, this and everything else is for Wendy, Alex, and Laura.

1
HOW
INTERNET
TELEPHONE
CALLS WORK

Why did Alexander Graham Bell waste time inventing the normal telephone and wires when he could have just invented the cell phone directly? Or why not just go ahead and invent Internet Telephony rather than just Telephony?

Our friend Alexander had to build upon the knowledge of his time (1876 for the telephone), as do all inventors. Telephony progressed quickly at the beginning but slowed when governments created monopoly telephone companies in order to spread the telephone as widely and quickly as possible. The government-granted telephone company monopoly (originally AT&T in the U.S.) was the trade-off for the push to put a telephone in every household.

Horrors

Can you imagine having only one telephone per household today? How twentieth century.

Universal Service was the term used in the past and is still used today to describe the goal of a phone for every household. In fact, your current telephone company still dings you a few cents each month to fund Universal Service, believe it or not. Although those few cents go to a noble cause, what about the other miscellaneous charges that somehow take a bill that starts out at, say, $11 a month, and bring it up to nearly $25?

One way to get around the ever-mounting charges by big telephone companies is to avoid buying services from those companies. Many people (millions and millions) are talking over the Internet for exactly that reason.

I know some of you are worried that something that uses the Internet must be difficult to install, configure, and use. Fear not. You're already using Internet Telephony–you just don't know it. You are also surrounded by ways you will use it in the future. Look at these details, gathered at the beginning of 2005:

- One-hundred and fifty million or more cell phone users in the U.S. connect over an Internet link for at least one leg of their call routes, even if they don't know it.
- Every IM (Instant Messaging) program that offers voice chat uses the Internet for this.

- More than a million game players on the Microsoft Xbox Live network can talk to each other using Internet Telephony while they play their games. And those users are scattered over 24 countries (*www.xbox.com*).
- Microsoft Windows XP (the dominant personal computer operating system) includes all the necessary software to support calls over the Internet.

Didn't know you were an experienced Internet Telephony user already, did you? Congratulations on your accomplishment.

Trust me when I tell you that this stuff is almost as easy as hooking up a DVD player, and certainly easier than configuring a computer. You can join the millions and millions of others who have preceded you into the future of telephones. Just like you, the majority jumped into this new world for one of two reasons:

- Free (or very cheap) phone calls
- Advanced telephone features impossible to get on traditional phones

You don't have to read this chapter to decide which Internet telephone service you prefer. You don't have to read this chapter to configure your chosen Internet telephone service. But I believe a basic understanding of what makes it all tick will make you a smarter customer and help you understand how calling over the Internet differs from the way you make calls now. This really is a different technology than landline phones and cell phones, with many advantages, but also with some differences that may seem odd at first.

I also think this is an interesting technology filled with clever people doing clever things. Keeping track of some factoids may even help you win a trivia contest someday.

Analog to Digital, Voice to Data

Analog means many things to different technical disciplines, but to me it means real life. Nature is analog; numbers are digital.

We speak in analog, because sound waves are continuously fluctuating and naturally created. Remember the graph of a sine wave you saw in school? Figure 1-1 shows how the sine wave looked then, and still looks today.

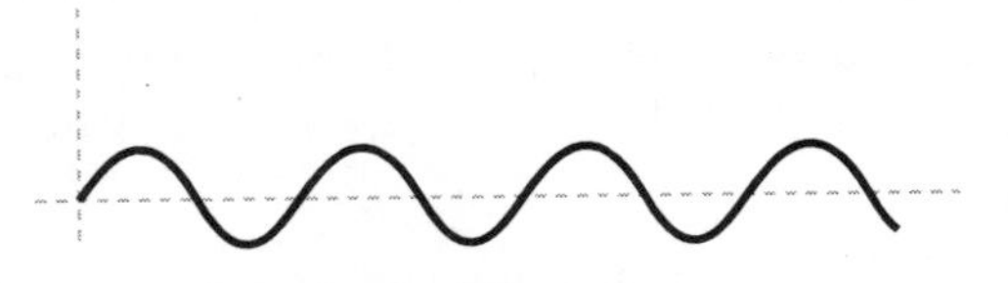

FIGURE 1-1. Your basic sine wave

Notice the wave form is smooth and continuously variable, such as what you get from a pure tone. You can illustrate the natural and continuous nature of analog yourself by singing a note and then sliding up and down through notes rather than changing from one note to another like a scale (you have my permission to wait until you're alone).

Analogies Abound

Tech factoid with literary hook: An analog device is *analogous* to the real world it represents, such as the hands sweeping over a round clock face representing the rotation of the globe (English teachers love analogies).

Telephone systems, starting with the first one built by Alexander Graham Bell up to the ones built a few decades ago, connected the speaker and receiver with copper wires. A diaphragm (microphone) in the receiver moved as sounds hit it, creating an electrical current. At the speaker, the electrical signals moved the speaker material to recreate the sound waves for the listener. Figure 1-2 shows a simplified telephone network connection.

Things are a bit more complicated, of course, but Figure 1-2 shows the analog world of early telephone usage. Every phone was physically connected to a telephone company central office, where all the switches connected your phone to the phone you called. There are central offices everywhere because they needed to be relatively close to the telephone at each end of the line.

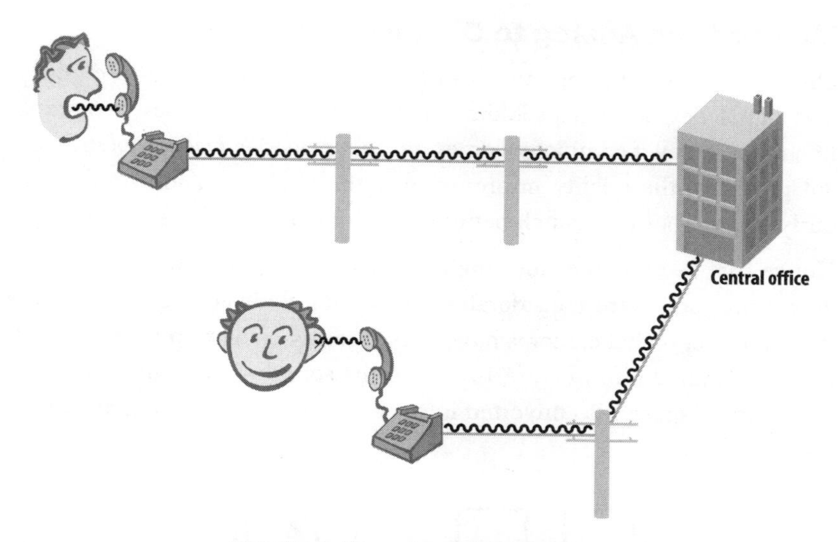

FIGURE 1-2. The analog world of telephony

If you're old enough, you may remember that long distance calls were scratchy, noisy, and difficult to hear. That's because amplifying (electrically boosting) the signal for long distances across multiple central offices always introduced noise into the circuit and degraded the quality of the sound.

If you don't remember long distance calls being of poor quality, then you didn't make any of those calls until the telephone company (AT&T) upgraded their systems and replaced copper wires for long connections with microwave, satellite, and fiber links. It's been decades since a long distance phone call spent the bulk of its journey over only physical wires.

Parts of your conversation that go over fiber cables, satellite links, and the like are digitized inside the telephone network and converted back to analog signals before reaching the other end. But the circuit for each voice call remains open, with all the *bandwidth* (some portion of the telephone carrier's overall capacity for calls and data, which is, unfortunately, a finite resource) in use that's needed for both people to talk at once, even though most of the time there's only silence on the line (between words, and of course the listener is silent while listening). All that silence wastes bandwidth, but the traditional phone call requires that enough bandwidth for both parties to talk at once be reserved–after all, if your phone didn't make it possible to have long distance shouting matches, what good is it?

Moving from Analog to Digital

One of the many reasons the world is moving from analog to digital is that signal degradation problem mentioned in the previous paragraph. Each time a new technology comes out that provides *higher resolution* digital products, the quality improves over the analog version used previously (remember how much better CDs sounded than cassette tapes?).

Notice that requirement for "high resolution" in digital products. Early digital products were considerably worse than their analog counterparts (for example, digital cameras have only recently caught up with the quality of traditional film). Figure 1-3 shows the sound wave in the sine wave shape from Figure 1-1 converted into a basic, low resolution digital form.

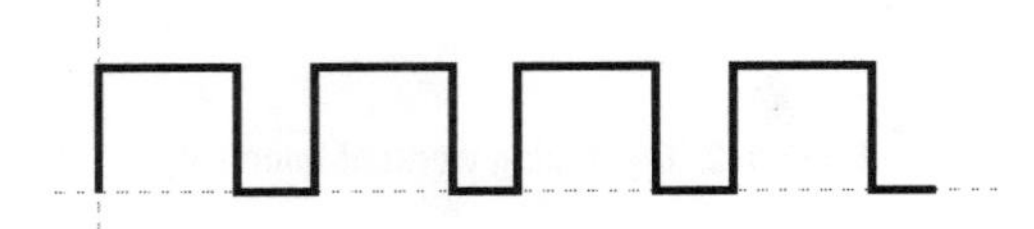

FIGURE 1-3. A basic digital signal

When I say basic, I mean basic. Digital information, popularly represented as 1s and 0s, really represents just on and off. That's it–on and off. Nothing in the digital world gets more basic than on or off, and that's the foundation of all digital technology.

A digital phone conversation at this resolution level hurts your ears (silence, beep, silence, beep etc.–speed it up and you've got the sounds of a dial-up modem, which is an analog representation of a digital signal). The digital representation of the sine wave in Figure 1-3 changes only state (moves from off to on to off to on) at the top and bottom of the sine wave signal. You either have silence or a beep.

As digital resolution improved, sound waves were measured more frequently. Now sounds are measured thousands of times per second (44,100 by standard Internet Telephony devices), and even millions of times per second in laboratory settings. Voice quality is no longer an issue, assuming there is enough bandwidth for the conversation.

One of the biggest advantages for digital over analog information is the ability to copy digital information exactly. Not copy information closely, but *exactly*.

Technology Secret Unveiled

Ever wonder why computer people often number the first of something a 0 (zero)? That comes from the digital *on* or *off state*. Off is the first state, or nothing (0), and on is something (1). That's why numbering and counting technology items sometimes start with 0 (for example, there are four plugs on the system: 0, 1, 2, and 3).

Yeah, that's dumb and should never have gotten out of the lab, but it did. Now you know.

A quick example? Let's say you write a joke (or steal it from the Internet and hope no one notices), print it out, and give it to two friends. Those friends Xerox that joke (excuse me for making that trademark a verb, they make a *xerographic copy*) and give it to two more friends. Those two copy their pages, and so on. After a while, the fifth-generation copies will be fuzzy and possibly difficult to read. But a copy printed out of your word processor will be just as clear as your original.

Ancient History

I wanted to use cassette tapes as the example, but my two teenagers weren't sure what cassettes were or why anyone would copy one.

Analog telephone conversations degrade when interference gets into the circuit anywhere along the line. Digital telephone conversations do not degrade, no matter how many switches they go through, because a digital message resists that interference. In fact, most switches clean up the messages by ensuring the timing spaces between bits remain uniform and signal levels remain correct.

Digital telephone conversations do have problems at times, but their problems are different. In fact, the problems are much more advanced and sophisticated and come from different areas than interference and noise in analog circuits. Luckily, digital telephone problems are rarer than analog telephone problems by a wide margin.

Analog to Digital Examples

Examples like these are good for parents or bosses who resist technology. Perhaps one of the examples will hit a soft spot and your technology resister will see that you're trying to help them with a Broadband Phone. With luck, they will then give you either a hug or a raise, depending.

Not every digital product has eliminated their corresponding analog ancestor, but new sales volume leans strongly toward the digital and away from the analog. And even in areas where that sales volume has yet to favor digital, digital products have the momentum, as shown in Table 1-1.

TABLE 1-1. Examples where digital products are replacing older analog products

Analog ancestor	Digital replacement
Clocks with hands	Digital clocks
Phonograph albums and cassette tapes	CDs
Movies on VHS tapes	Movies on DVDs
Cameras with film	Digital cameras
Traditional radio broadcasts	Satellite and Internet radio

This table doesn't proclaim that every digital technology automatically renders the earlier analog technology obsolete and useless. Personally, I prefer clocks with hands that show the sweep and continuation of time, but that philosophical argument annoys my teenagers. Until recently, the resolution of film in cameras was higher than that of digital, but in the most popular size, 35mm, digital cameras now have more resolution than the film cameras they replace. That is not yet true for larger film formats.

See, digital phone conversations are a natural evolution as digital technology develops to the point that it can equal and even improve upon analog phone conversations. While some of the phone technology and business models in Internet Telephony can rightly be called revolutionary, the idea that telephones would go digital is an evolution of society's constant improvement.

VoIP (Voice over Internet Protocol)

Let me say, first off, that VoIP is a terrible choice for an acronym, which degrades the technology and benefits provided. If it was called it CRAP (Conversation Remote Access Protocol), at least it would be funny and middle school boys would get a big laugh. But we're saddled with VoIP (say "toy: with an ending "p" then replace the "t" with a "v"), a graceless sound that calls to mind a love-starved Amazonian parrot's mating call.

Name Chain

Internet Telephony = VoIP = digital phone conversations = less dependence on the traditional telephone companies = savings and more features.

That said, let me offer you three levels of explanations: Overview, Some Details, and More Details. When you read enough details to feel comfortable that the technology works generally and will work for you specifically, skip the rest and go to something else in the book that interests you. If you want details, keep reading.

Minimum Requirements

There are three things you must have to use *traditional* phone services:

- A telephone line
- A phone
- An account with a phone company

There are four things you must have to use Internet Telephony (I won't ever say that four-letter word–VoIP–unless absolutely necessary). Here are the minimum requirements for talking over the Internet:

A broadband connection

A must-have, and don't believe the few companies who say you can use a standard telephone line and a modem. Cable and DSL broadband customers will have no problems, and are already inundated by sales pitches from their providers. Satellite broadband users must have satellite service that operates at broadband speeds both ways, and even then you shouldn't bother because you won't be happy with the delays introduced by the satellite distances.

A telephone or equivalent

You can get adapters to use a standard telephone with an Internet Telephony service. You can also use a microphone and your computer's speakers, or a headset that plugs into your computer in place of a telephone. But you need something to talk into and hear out of.

A computer

Softphone systems use special software to turn your computer into a telephone, so obviously, you need a computer. Internet Telephony services normally provide an adapter to connect your existing telephones to the Internet for calling, but you will need a computer to configure and monitor your service. Special Internet telephones have a small computer inside them, and for the services that work on handhelds (such as Skype for Pocket PC)—well, your handheld is the computer.

Electrical power for the telephone

Traditional telephones are powered over the phone lines by the telephone company, but Internet telephones need power from some other source.

Power Reserves

OK, that last item may have given you pause. What happens if the power goes out? Watch for details in Chapter 4.

Notice what's missing? An account with any type of phone company. You can bypass that if you want, especially the part where you send them money each month. Listing yourself with a service (as you do with Skype) so others can find you is somewhat like an account with a phone company, but it's more like registering your address than subscribing.

Internet Telephony services do act more like a phone company by switching your calls and providing services, but they aren't traditional phone companies. They will charge you a monthly fee, but that fee includes far more services than you can get with a regular telephone and also reduces costs for long distance calls by a huge margin.

Pick A Name

Since others also avoid using the term VoIP, a variety of names have been created. Internet Telephony is a good one, as well as Broadband Phones or IP (Internet Protocol) Phones. One of these days we'll just call them phones, just as we now call cordless phones just phones.

There are a variety of other things you may want to use for added features and flexibility, but the minimum requirements are pretty minimal. Well over 30 million homes in the U.S. already have broadband, and that number is expected to reach nearly 50 million by 2009, so the market for Internet Telephony is a large one that continues to grow.

Basic Internet Telephony Overview

Internet Telephony converts your voice into *data packets*, similar to what the traditional phone company does when sending voice over fiber cables for long distances. At the far end, the data packets are converted back into sound signals and the voice you hear.

Data in Transit

A packet is a block of data. A block of data must almost always be broken into small pieces for easier transmission. Packet sizes are pretty small, ranging from 53 bytes up to around 1,500 bytes (in a plain, unformatted email message, a *byte* corresponds to a single letter, digit, or symbol) depending on the specifications of the network carrying the packets.

Think of 500 tourists all going to the same cathedral. No one vehicle will carry them all, so they split up. Some may go in groups of 4 in cars and some may go in groups of 40 in buses. Data packets work the same way, but they're usually better dressed than tourists.

The main difference isn't that voice sound waves are digitized for easier transmission, but that the Internet, rather than the telephone company's network, carries the signals. But this is a big change in how voice calls, the link of one person in one place to another person in another place, are handled.

While the traditional phone system connects to a phone number attached to a physical device (desk phone, cordless phone, or cell phone), Internet Telephony can route a phone connection to a variety of network devices. This can be done because the Internet telephone call is treated like data, and can be routed to a device based on where a user is logged in. If you've used Instant Messaging from both home and work, you know that people can find you when you're logged in from either location, and have experienced this already. Table 1-2 shows some of the key differences and similarities.

TABLE 1-2. Telephone calls over the traditional telephone network and the Internet

Traditional telephone	Internet telephony
Dial a regular telephone	Dial a regular telephone
Connect to your central office	Connect to your Internet Service Provider
Call goes over the telephone network	Call goes over the Internet
Call connects to central office at far end	Call connects from Internet to central office on the far end
Phone rings at the far end	Phone rings at the far end

Check Chapter 2 for details about which type of Internet Telephony service will work best for you, but you can easily replicate your existing telephone experience using a broadband phone. In fact, you can completely replace your existing telephone, including services such as 911 calls, with Internet Telephony. You will get similar services for similar money, and Internet Telephony will give you more features. People can call you, and you can call anyone in the world with a phone, whether they have Internet telephone service, a cell, or an old-fashioned phone.

More Internet Telephony Detail

Although the process of turning analog voice streams to digital packets is similar to that already used by the telephone companies, there are differences. The traditional telephone companies are still bound by the demands of circuit switching and the need to keep a circuit continuously

network connections. The goal of the Internet is to deliver every packet regardless of where it came from, where it's going, or what's in that packet. Packet in, packet routed, packet delivered, billions and billions of times per hour, no matter if parts of the network are congested or offline. Packets in, packets routed, packets delivered.

When intelligence is expensive, such as nearly 100 years ago when AT&T started building the network, centralizing the intelligence makes sense. But now intelligence, in the form of processing power, is cheap and portable. Somewhere, no doubt, someone has measured the processing power of a Palm Pilot against the collective intelligence of all the switches in New York City in 1912 or some such.

The point is clear: intelligence and processing power is now extremely inexpensive, and the capabilities of the basic telephone connected to the telephone network was long ago surpassed by intelligent devices attached to the Internet. The smarter the end points, the dumber the network can be.

And the Internet was designed to be dumb and focus on moving bits and nothing else. When a router (a device used by ISPs and others to make sure data goes where it's supposed to go) on the Internet goes down, data streams flow around that router using different tracks through the Internet and still reach their destination. Once you get past the link from your home computer to your broadband provider, there is no other single point of failure on the Internet. Until, of course, you reach the last leg down to a home computer on the far end.

The last segment, between your provider and you, is called the *last mile* of a connection. No, it's not always a mile, but that's the term. Who owns that last mile? In almost every case, one of two companies provides the last mile: your telephone company or your cable provider. Many millions of homes have both connected, giving them, I suppose, two last miles.

On the Internet, the smart end devices connect over a dumb network. On the telephone network, dumb devices connect over a smart network. Fans of both will argue their way is better. However, it's much easier to upgrade a few end devices at a time to add new capabilities than it is to upgrade the smarts of an entire network. Upgrading the end devices is what Internet Telephony is all about.

Large Internet service providers make arrangements with each other, called *peering arrangements*, to connect their networks directly and keep their traffic off the public Internet. After all, if they keep their customers' traffic on private data channels, they have more control. Internet Telephony companies are starting to work out voice-peering arrangements for exactly the same reason. Expect service to have higher reliability and the vendors to have more control over the voice traffic flow as more peering arrangements are signed. Yes, this is an example of more innovation from the Internet Telephony companies.

Bits Are Bits

Once the audio portion of a telephone call has been digitized, whether 8,000 times per second or at a higher resolution, the audio becomes a stream of digital bits. These bits are virtually indistinguishable from bits of web sites, music files, email, and airline ticket confirmation forms. Bits are bits.

With a monolithic telephone company network, timing must be tightly controlled so connections between central offices tick the same seconds off together. But the decentralized Internet has a wide variety of timekeepers, and few networks absolutely track connected networks to the subsecond level. Yet Internet Telephony works fine.

The data stream packets require little absolute timing, but do require coordinated timing between packets in the data stream. So the Real Time Protocol portions of each data packet help coordinate packet sequencing at the receiving end, keeping words flowing smoothly.

Courts have followed the "common carrier" rules for Internet service providers, meaning they are not responsible for what goes over their lines. After all, the Post Office isn't held responsible for what individuals mail, and the telephone companies have never been held responsible for what people say to each other over the phone. If the people mail or say something illegal, they may get caught, but the Post Office and telephone company are not considered coconspirators. Bits are bits.

And bits in the late 1970s were bits, at least to a study by the Department of Defense. Their own widely publicized report said clearly that a packet switched network, supporting packetized voice traffic, would be cheaper than circuit switched voice, the only technology in place at that time. Bits are bits, and 30 years ago, bits were bits.

Regulatory Issues

We can't blame AT&T engineers because they didn't reinvent the telephone on a regular basis; that wasn't their job. The U.S. Government made a deal with AT&T to grant them a monopoly in exchange for building the world's best telephone system. AT&T did their job, but politics and innovation changed the rules on the old guard phone company.

But regulation remains a popular issue for some states and some federal lawmakers. They believe that because telephone calls are regulated, Internet Telephony must also be regulated. However, cooler heads believe that a national telephone network is regulated, not the telephone calls themselves, and packetized voice across independent data networks (the Internet and private networks) should not be regulated.

Here's a clip from the FCC web site (*www.fcc.gov/voip/*):

> Does the FCC Regulate VoIP?
>
> The Federal Communications Commission (FCC) has worked to create an environment promoting competition and innovation to benefit consumers. Historically, the FCC has not regulated the Internet or the services provided over it. On February 12, 2004, the FCC found that an entirely Internet-based VoIP service was an unregulated information service. On the same day, the FCC began a broader proceeding to examine what its role should be in this new environment of increased consumer choice and what it can best do to meet its role of safeguarding the public interest.

In spite of their insistence on bleating the graceless word *VoIP* at me, I love this paragraph.

Thank You, Founders

The Founding Fathers mandated that all government information (with very few exceptions) may be freely copied, quoted, and used. Copyrights do not apply to government-created information.

Of course, this all gets tricky when you realize that the telephone companies, which are heavily regulated, provide many of the Internet data connections used by all the people talking over the Internet and bypassing the telephone network. Worse, when an Internet phone call goes to a phone that's part of the traditional telephone network, an Internet-to-PSTN gateway must connect the Internet phone service to the traditional telephone service.

This overlap confuses some people, and they demand regulation. After all, they regulate AT&T, and they want to regulate the Internet telephony calls running over AT&T's Internet data connections.

Many experts feel that the reason the Internet Telephony should not be regulated in any way is because the AT&T data lines are just carrying bits. Moreover, customers leased those AT&T lines and their bandwidth to support any and all types of Internet traffic. No one leases a data link just for web pages, and a separate data link just for email. You lease a data link and you control the information flowing through that data link, period.

The move from regulation to free market always hurts the entrenched status quo industry leaders, and the telephone business will not escape this pain. Telephone companies will shrink and be forced to innovate for their own survival as they lose customers who switch to less expensive options.

Regulation has yet to help the incumbent telephone companies, even though they go crying to the government every chance they get. Jeffrey Citron, CEO of Vonage, Inc., told me the incumbent phone companies "proved they will abuse their market power to keep their high pricing. They're not acting in the best interests of their customers, so that's why they're heavily regulated. If they want to get out of their regulated business, they can build out the fiber to the home networks they promised, because those are nonregulated businesses for them."

Every step of the way, incumbent phone companies fought Internet Telephony proponents rather than innovating new technology. In fact, the standard line is that while Internet phone companies hired engineers, the remnants of the original AT&T network hired lawyers.

Some Internet Telephony executives expect the incumbent traditional telephone companies to drop traditional wire-line telephone customers. You may think this is just wishful thinking, but AT&T, of all companies, has declared their intention to drop out of the residential telephone long distance service market, and has dropped in value so much that one of the Baby Bell spin-off companies, SBC, bought them. Another major telephone company is trying to sell 30 million home telephone customers to anyone who will take over their landline business.

Redial

Internet Telephony is a disruptive technology for several reasons. First, it sprang up without government funding and restrictions. Second, for the most part, it bypasses the strong traditional telephone companies completely. Third, the pricing of Internet Telephony can be so cheap as to be free. In fact, when you pay the extra cost to get broadband service, you might start demanding several Internet Telephony company options as part of the package.

Everyone (Internet companies, startups, existing cable companies) are attacking the traditional voice telephone market using packetized voice connections over the Internet. You're getting in on this technology wave early, but you'll have plenty of company over the next three years. In five years, someone not talking over the Internet will be the old fogies, and all the cool people (you and I and millions of our friends) will be using the Internet to prove every day that Talk Is Cheap.

2
YOUR
INTERNET
PHONE

All telephones and services making calls across the Internet are not the same. Huge differences exist in the features available to you depending on the type of service you get.

There are no adequate general terms used to describe the two different Internet Telephony approaches available today. So I decided to make up my own terms:

Phone-centric

This phone service duplicates the experience of traditional phone service you are used to with your traditional phone, and focuses on the phone, since that's what you touch and talk into. You can use your current telephone if you get an adapter that plugs your telephone into your broadband service. You may even be able to keep your existing telephone numbers. The leading company following this model is Vonage.

Computer-centric

This phone service creates a new method of talking between two people and bypasses the traditional telephone companies almost completely. Your contact point is the computer, either with a headset or a telephone that plugs into your computer. The leading company following this model is Skype.

Table 2-1 shows a quick comparison between the two types of service available. Generally, phone-centric looks much more like the telephone system you're used to, while computer-centric offers lower cost and new features for users who rely most on their computer (desktop or portable).

TABLE 2-1. Capabilities of the different Internet phone systems

Capability	Phone-centric	Computer-centric
Free calls to other users who are using the same service	Yes	Yes
Call traditional telephone numbers	Yes	Optional
Receive calls from traditional telephones	Yes	Maybe
Caller ID (show who's calling)	Yes	Yes
Three-way or conference calling	Yes	Yes
Voicemail	Yes	Optional
File transfers	No	Yes
Automatic call encryption	No	Yes

TABLE 2-1. Capabilities of the different Internet phone systems (continued)

Capability	Phone-centric	Computer-centric
Call from anywhere with a broadband connection	Yes	Yes
Built-in Instant Messaging and file transfers	No	Maybe
Missed call notification	Yes	Yes
Call forwarding	Yes	Optional
Choose the area code for your broadband phone	Yes	Optional
Separate fax line	Optional	No
Make and receive calls from wireless laptop	Yes	Yes
Make and receive calls from wireless PDA	No	Yes

When you check the various broadband phone providers (there are nearly 500), you will read an enormous amount of overactive marketing materials (okay, hype). Seemingly every phone service provider calls themselves a revolutionary breakthrough in the history of human communications. They are wrong.

Phone-centric service providers look quite a bit like traditional phone companies. They have centralized systems, like the old telephone company switches, but they use data-networking routers to transfer calls rather than huge traditional telephone switches. They use broadband (the coaxial wire used by cable providers or the telephone lines used by DSL providers) wires for their phone connection rather than the two-pair copper wire used by the traditional phone companies, but it's still wire. When providers start running fiber optic cable to each home for truly high bandwidth broadband service so they can sell us more services like video on demand, it's still a physical wire coming into the home. These companies claim they revolutionize the telephone, but they are only an improvement, not a revolution, in the technical world.

Acronym Alert

DSL = Digital Subscriber Line, the telephone company's broadband technology.

Computer-centric service providers looked at the technology from a new angle: put all the intelligence in the end devices (computers of one kind or another rather than telephone handsets) and let each caller connect directly to their desired callee over the Internet. In other words, they have the minimal amount possible of centralized services (a listing of other users using authentication software to track them when they log into the service) while doing everything possible on a peer-to-peer basis. These companies claim they revolutionize voice communications, and they have reworked the traditional telephone model into something new by focusing on the voice part of the equation rather than the hardware part. They may be right.

Revolution, Evolution

In marketing, your company is always revolutionary while your competitors are always evolutionary. For example, your competitor is evolutionary because they slapped a new coat of paint on their product, but you are revolutionary because you redefined technology through color differentiation. Yes, that's a new coat of paint on the product, spun by marketing folks.

One way to illustrate the two major different approaches concerns the focus of each technology. Phone-centric services put the focus on the telephone and on re-creating your traditional phone experience as much as possible. Computer-centric services focus on the computer as a means to transmit voice, as well as other data. Unfortunately, the dividing lines go out of focus on a regular basis, and will continue to do so as new products enter the market with new features.

Drilling down a little, let me show you three different ways that combine the two major methods with a few twists here and there:

Any phone to any phone

This is the phone-centric answer and the one most comfortable to most consumers. Your existing phone plugs into an adapter that connects to your broadband service, rather than into the old telephone jack. Modern data switches at your new telephone provider connect your call to any phone connected to the traditional phone company (including cell phones) and any phone connected to another phone-centric provider. The reverse is also true, meaning you have a phone number reachable by every phone in the world. But talking over the Internet rather than the old phone switches keeps the price low and adds a bunch of modern features. This is the most expensive type of Internet Telephony (but still lots cheaper than traditional phone service), and the major player in this market is Vonage.

Computer software phone to any phone

A headset connected to your computer makes calls through your broadband provider to a third party's data center (such as Skype's) and on to the Plain Old Telephone System central office nearest the number you dialed. These calls are less expensive than regular long distance because the Internet carries the call for the longest stretch and then just makes a local call at the far end. Depending on who you've signed up with, other people may be able to call you from any phone. The major player in this market is Skype with their optional SkypeOut feature (but other people can't call you; you can only call out with SkypeOut).

Computer software phone to another computer software phone

This is the earliest type of Internet Telephony that completely bypasses the traditional telephone company network. Only computers with the right software can connect to each other, and there is no phone number available for outsiders to call you from a traditional phone. This type of service is almost always free, but limited. Skype is the leading player in this market, but there are scores of others with similar features (instant messenger programs have had voice chat in some form for years). Unfortunately, you often can't call from a computer using one brand of software to another computer using a different brand of phone software.

Confusing? A little bit, but the consumer market for broadband phone calls still needs to mature. Some of the "gee-whiz" announcements flying out of small companies today may never become real products or services. Many of the companies making the most noise today will be bought by older, more established telephone companies tomorrow. But the technologies have pretty well stabilized (even if some of the implementation details haven't), so you can make a choice today that will still provide excellent service next week and next year. And if you choose

Too Many Choices?

Companies announce new twists on these major options seemingly every day. Sometimes the twist is to put the computer software into a special phone (often called a digital phone), and sometimes the twist is to divert your existing phone line so that it connects to a broadband provider's office instead of the phone company's central office (they convert the analog phone signal into a digital data stream at their office rather than using special equipment in your house). And sometimes your digital phone is a combination of a wireless network data connection and a traditional cell phone, letting you choose which service you prefer if more than one is available at your location.

wrong, number portability will let you change services and keep your number in most cases, making it simple to recover and start afresh with a new service provider.

Phone-Centric Providers

The companies following the phone-centric method are the closest to the traditional telephone companies. For many people, the most important thing is that any phone using their service can call any traditional telephone and cell phone. Even more important for many is that any traditional telephone can call a user on these services just like they can call any other telephone. This type of service will appeal to many traditional telephone company users who are looking for more features, more savings, or both.

Old Joke

The old joke about "even your mother can handle this technology" is absolutely true.

Phone-centric companies feel much like traditional telephone companies to the users. The major differences under the covers are:

- They use broadband connections rather than telephone lines.
- Their routers and switches route calls over the Internet until they reach their data and switching center, then they pass the call off to the telephone company central office close to the traditional phone being called.

The first point tells us that the traditional telephone company's pair of copper wires hanging from a pole behind the house and looped over to the roof is no longer necessary. In the past, there wasn't enough technology available to deliver more than basic voice service over a single pair of wires. Now technology allows companies to deliver telephone and TV services over a single coaxial cable installed by the cable company. Time and technology marches on, and the telephone companies are busy providing DSL broadband service (including TV in some test programs) over those same old copper wires that were able to provide only voice calls a few years ago.

The second point tells us that the huge telephone network built by AT&T to handle voice traffic has become, in many cases, redundant, if not downright unnecessary. Our history lesson in Chapter 1 told us how the government granted AT&T a monopoly in order to pay the huge expense necessary to get telephone service to every household (called Universal Service, which you're still paying for every month). AT&T did a great job of building that huge wired network across the country. But Henry Ford did a great job building Model Ts, and we don't need those anymore, either.

The old copper wires remain critical in reaching homes and businesses without broadband connections. Every phone still part of POTS (Plain Old Telephone System) connects to the rest of the world through those two copper wires strung by AT&T so many years ago. Of course, when everyone gets some type of broadband connection to their home or business, the old copper wires can be retired.

Here are the good points about phone-centric service providers:

- You can use your existing telephone.
- They provide, for free (or for a small setup fee), a device that connects your telephone to your broadband service.

- Other people can call you just as they do now.
- In most areas, you can keep your existing telephone number.
- 911 emergency services are available in many areas.
- Sound quality is as good or better than your traditional telephone service.
- The telephone adapters provided by many services will enable you to connect several computers to your broadband connection, often using wireless networking.

These items are separate from the advantages Internet Telephony provides by including all number of features called "advanced" by the traditional telephone companies (they charge "advanced" prices as well). Those appear in Chapter 3, with explanations and the amount of money you're saving by getting those features with a broadband phone rather than a traditional phone.

Yet paradise has not been completely attained. There are some issues to consider when moving away from your traditional telephone service. These considerations include:

- You need electrical power to make phone calls, where traditional telephone service sends enough voltage down the phone lines to power basic telephones.
- Your phone extensions must be plugged into the broadband service adapter, rather than your existing phone jacks in the wall (for most services–several vendors are working on eliminating this situation).
- Cable and DSL network connections do not have the same high rate of uptime as the traditional telephone network, but they are getting closer every day.
- Not all providers can get your phone number put into the local phone directory (but you could consider a free unlisted number as another feature).
- You may not be able to receive collect calls or dial pay-to-talk numbers.

Can these situations be handled without undue hassle? Absolutely. For instance, the situation requiring electrical power that is provided by the traditional telephone company may sound difficult. However, this requirement is no different from using cordless phones, which need power, and millions and millions of people use those happily. But now

A Wide Enough Pipe

A most significant consideration is that you will need at least 90 kbps upstream bandwidth from your broadband connection to be happy with the voice quality (low-end DSL provides only 128 kbps upstream bandwidth), and that speed rating doesn't leave much capacity for other upstream traffic (outgoing email, peer-to-peer file-sharing applications, and so forth) during a call without compromising the voice quality. If you don't know what your upstream speed is, you should contact your broadband service provider. They will also be able to let you know how much it will cost to upgrade to a faster speed–if you find that your Internet connections slow to a crawl when someone is using the phone, then it's likely you need a faster connection.

you can buy a battery backup, just like for your computers, to power your broadband phone equipment. The situation with the phone extensions is a little more problematic. Most homes have telephone connections wired to nearly all the rooms, especially if the home was built in the last 15–20 years. Even many bathrooms have phone connections.

If you work with a home entertainment consultant or broadband phone company catering to business users, you can tie a broadband telephone connection into the existing home wiring. That will cost more money than many people wish to spend, however.

You can bang your head against this wall, or you can get a workaround. I suggest you save the wear and tear on your forehead and buy a new cordless phone with multiple wireless extensions. Put one extension in every room you have a phone today. Problem solved. Details await in Chapter 4.

In fact, your problem's solution will improve your telephone life. Cordless phones are portable, intelligent (lots of memorized numbers and call-handling features), usually come with Caller ID support, and provide intercoms between rooms in many cases. You will spend somewhere between $100 and $400 depending on how many phones you want (you can buy up to eight extensions with many vendor's phones), but you will wind up with more control over your telephone than you have now.

You can also choose phones from major vendors (such as Motorola) that connect extensions via your home's electrical wiring. Since you have to plug a cordless phone into the wall anyway (guess we should only call them half-cordless), take advantage of well-proven home-wiring networking if you prefer that over a cordless phone option.

Another choice is to add your broadband phone as a second line at your desk by your computer and broadband network connection. This option keeps your existing phone and adds a broadband phone for all the long distance calls and other new features.

Vonage

Technically called Vonage Holdings Corporation, Vonage (*www.vonage.com*) leads the phone-centric marketplace through three things: their early arrival in the market, well-greased wheels provided by venture capital money, and aggressive marketing. They remain privately held, well-funded, well-established, and they keep a high profile with constant media appearances by Jeffrey Citron, their CEO. How aggressive is Jeffrey? When I contacted Vonage about information for this book, Jeffrey made an appointment to talk to me before I could talk to any of their technical or marketing people.

Vonage was founded in January 2001, in Edison, New Jersey. AT&T had their Bell Labs in New Jersey for decades, so new phone companies seem drawn to that area. Perhaps, as all roads led to Rome in the old days, all phone lines run mystically to New Jersey today.

The company has over 600 employees, meaning they are substantial. That doesn't guarantee they will succeed, but it means they have some current traction.

Over 400,000 phone lines are active with Vonage at the beginning of 2005, and they reached 500,000 users by springtime, meaning they have a large subscriber base and some regular revenue. Building revenue proved extremely difficult for some startups in the Internet space during the boom years–when the bubble burst in 2000, those companies disappeared. Vonage appears to be large enough not to disappear, but not large enough to fight off a takeover by some of the huge players in this marketplace.

Vonage has over two dozen regional data center locations spread across the United States. They have a growing handful in other countries (two in Canada, one each in Mexico and England). There are well over 2,000 locations served by the full complement of Vonage phone services across the U.S. and a few other countries. Outbound calls from any Vonage customer can reach any traditional telephone in the world, whether Vonage has clients in that country or not.

Executives in the company come from a variety of well-heeled former jobs, including a fair number of successful startups. This company is not a guerilla phone company; it is very corporate, and it hopes to become the next version of a modern telephone company while putting former AT&T companies out of business.

Calling them "brash" fits, as does "so confident as to be almost arrogant." Since the company remains private at this writing, they don't have to admit how much venture capital money they have received (a lot, up into the $400 million range according to some estimates) or how much of that money they have burned through (much of it, according to detractors and/or jealous competitors). Vonage is high profile on purpose and enjoys the spotlight. I believe they have a good chance of remaining on top of this market.

Other Internet Telephone Service Companies

The complete broadband telephone service model requires some serious investment by the vendor, but that hasn't stopped nearly 50 companies from jumping into the market. The reason you have heard of only a handful of providers is because many are regional resellers of a national service. Of course, it also takes plenty of advertising and marketing dollars to become known, and fewer than a dozen have made any national market penetration at all. Do not take the list in Table 2-2 as definitive, because it will change on a regular basis.

Prices will also change (mostly downward if the normal trend holds), so price lists will almost assuredly be out of date whenever you read this book. I will list company prices as low (under $20 a month), average ($20–$25 a month), or high (over $25 a month) as a point of comparison.

Each of the companies listed in Table 2-2 include some type of telephone-to-broadband adapter to allow you to keep using your existing telephone. These companies give you a new telephone number or allow you to keep your current number, and anyone using a traditional telephone can call you.

TABLE 2-2. Broadband phone companies

Company	Price	Feature set	Market penetration
Vonage *www.vonage.com*	Medium	Large and complete, including 911	Market leader
AT&T CallVantage *www.usa.att.com/callvantage*	High	Complete, including 911	Medium
Time Warner *www.timewarnercable.com/corporate/products/digitalphone*	Very high	Complete, including 911	Medium
Verizon *www.voicewing.com*	High	Average, no 911	Low
VoicePulse *www.voicepulse.com*	Medium	Complete, with interesting additions, no 911	Low
Packet8 *www.packet8.net*	Low	Average, fee for 911	Growing
BroadVoice *www.broadvoice.com*	Low	Average, no 911	Growing
Lingo *www.lingo.com*	Low	Average to good	Growing
Net2Phone *http://web.net2phone.com/consumer/voiceline*	High	Average, no 911	Low
RyanTech *https://voip.ryantechinc.com/index02.php*	High	Average, no 911	Low
MyPhoneCompany *www.myphonecompany.com*	Average	Complete, no 911	Low
VoxFlow *www.voxflow.com*	High	Average, no 911	Low

TABLE 2-2. Broadband phone companies (continued)

Company	Price	Feature set	Market penetration
OptimumVoice *www.optimumvoice.com*	Very high	Average, advanced wiring options, 911 included	Low, geographically limited
SBC *www.sbc.com*	Not set	Not set	Available mid-2005[a]
Broadvox *www.broadvox.com*	Not set	Not set	Available mid-2005

[a] This changed when they bought AT&T and took control of their AT&T CallVantage service.

This market is heating up considerably in 2005, and companies will add features and drop prices to respond to competition. Check carefully when looking to start your service.

However, once you have a service, you won't gain much by switching every time a competitor drops their price a dollar or two. Saving money is a good thing, of course, but reliability, support, and an established track record of good service by your provider is worth a couple of dollars a month. Every time you switch a service, you have to relearn some of the procedures and run the risk of being without service for a couple of days (or more) during the transition.

Computer-Centric Providers

Every computer attached to the Internet (with the right software) can communicate with any other computer attached to the Internet. We saw in Chapter 1 that every computer sold for the last decade has the horsepower necessary to turn voice into digital data streams. Put those two facts together and you get a revolutionary jump in communications technology: any computer can send and receive voice streams to any other computer across the Internet.

Peer-to-peer just means that two devices connect directly to each other without a server managing the transaction (and that's ignoring all the routers and other Internet servers handling data traffic). When you physically talk to a person in the same room, it's peer-to-peer. When you mail that person a letter, the Post Office acts as a server in the middle, so it's not peer-to-peer. When you call that person on the phone, the telephone company and central office switches get in the middle, so it's not peer-to-peer.

The most used peer-to-peer connections today are web clients to web servers. When you user your browser to view a web site, the systems connect directly to each other without any server in the middle handling the transaction.

Some type of directory service must be available to help users connect to each other. Every device that can be seen publicly on the Internet has a unique sequence of numbers called an IP (Internet Protocol) address. Multiple types of directory services help decipher a long number (such as 65.254.50.98) to an easily remembered web site name (*www.gaskin.com*).

Brick Up That Wall

Even computers that aren't publicly accessible, such as your home or office computers (which should be sitting behind a protective firewall), have IP addresses. The difference is that those IP addresses are not unique (the computers in many other offices probably use the same addresses). What *is* unique is the IP address of your broadband router, and it's capable of performing a number of tricks that make it possible for computers to talk to one another through the firewall.

Notice what is missing: the web page equivalent of the telephone company's central office switches. A directory service is a software database and is optional. After all, if you know the phone number you want to call on your cell phone, you don't need a telephone directory, do you? Directories help peer-to-peer users find each other, and are not mandatory. But they are very handy.

Skype

Enter Skype, developed by the two programmers who founded the file-sharing program KaZaA (Niklas Zennström and Janus Friis); Skype hopes to connect everyone via their high quality and free software and telephony service.

The two Skype founders have since sold KaZaA, so don't yell at them if you believe sharing music files is illegal or if you feel music should be free. But they learned many things from KaZaA that they applied to Skype, including how to create a free program so exciting that people spread the word around the world on their behalf. Experts call this *viral marketing*, since it spreads like a virus from person to person. (That's a compliment, really, although if you think about it too long, it sounds disgusting.) Just remember that in marketing, brand awareness wins, and having every user of your product (or free Skype software in this case) demand that all their friends get the product is a wonderful thing and much desired by marketing people.

Name from Air

"Skype" is a made-up word picked because it had an available web address, the ability to act as a noun and a verb, and no meaning in any language.

Peer-to-peer technologies that were developed for KaZaA and improved for Skype allow the two founders to claim that Skype is the third generation of peer-to-peer technology. That makes sense, and explains why Skype has an advantage over many of their competitors: they've been doing it longer and learning as they go.

Starting August 29, 2003, Skype software rolled quickly across the Internet. People were so excited that they gladly suffered through the early growing pains. Of course, a cult-like following was all they had, because Skype users could only talk to other Skype users, and only when they were both on their computers and logged into the Skype service.

The word spread, and by early 2005, the Skype software was downloaded well over 75,000,000 times, and 1.3 to 2.1 million users were online at any time of the day. 10,000 new users register per day, adding up to over 21 million unique usernames listed in the Skype directory service. It appears that viral marketing worked well in this case.

Skype folks are quick to say there is no central server running the service, exactly what you would imagine for a peer-to-peer network system. But they do have plenty of servers in place to run their SkypeOut service, which connects a Skype user to the POTS (Plain Old Telephone System) to reach non-Skype users.

There's more about SkypeOut and all other things Skype in Chapter 6. You'll also learn how a company giving away free software hopes to make money, even when people use the product for free over the Internet.

Other Peer-to-Peer Telephone Providers

Startup costs for a peer-to-peer telephone "company" are pretty low: software. In fact, you can get open source software free for the downloading (*www.sipfoundry.org*) that offers a complete softphone. Pay a programmer to customize the look and put your name on the softphone, or do it yourself, and you're in "business" as a provider of Internet Telephony.

Soft, Not Real

A *softphone* is software used to emulate a real telephone through computer software. You will need a headset plugged into your computer to use a softphone successfully.

Whenever you see a company offering software to download to use Internet Telephony, as Skype does, you will probably get a softphone. Xten (*www.xten.com*) has a polished commercial version of a softphone, as well as a "lite" version available for free. Figure 2-1 shows their softphone being used by an Earthlink customer.

You can see the number pad, the volume settings for both microphone (left) and speaker (right). The on and off hook icons are on the left and right as well.

What you *don't* see is the computer running this application and the headset plugged into that computer. I do like the way Xten put their little marks at the top (speaker) and bottom (microphone) of the softphone to make it look like a fat cell phone. Notice the little circle on the bottom right? Looks like the earpiece plug on your cell phone, doesn't it? An excellent way to bridge the gap from the familiar (cordless or cell phone handset) to the new (softphone), isn't it?

FIGURE 2-1. A free softphone from Xten.com

By Any Other Name

Vonage offers a softphone option with their service. If you go to their web site, you'll see this same Xten softphone with Vonage's name on it. But their photo wasn't as good as this one, which I got directly from the Xten folks.

Free World Dialup (*www.freeworlddialup.com*) is a peer-to-peer software telephone company that actually appeared before Skype. They have a softphone background, but are moving beyond that, as has Skype.

There are dozens of other softphone peer-to-peer telephone companies available across the Internet. Your Internet Service Provider may offer this service.

Many businesses use these types of phones to communicate between offices and individuals, especially in the technical fields. After all, if your computer is always on when you're working, it can work just as well as a "normal" telephone. And because peer-to-peer telephone calls across the Internet are free from every peer-to-peer service, those wanting to avoid long distance charges can do so easily. Just remember that Internet-only telephone companies have a severe limitation: talking to the non-Internet telephone user.

Peer-to-Peer to Public Telephone Network Connections

Unfortunately for even the hardcore technogeek, we all have friends who have yet to jump on the Internet phone bandwagon. Therefore, even the free-est of free Internet telephone services must have some way to connect to the rest of the world.

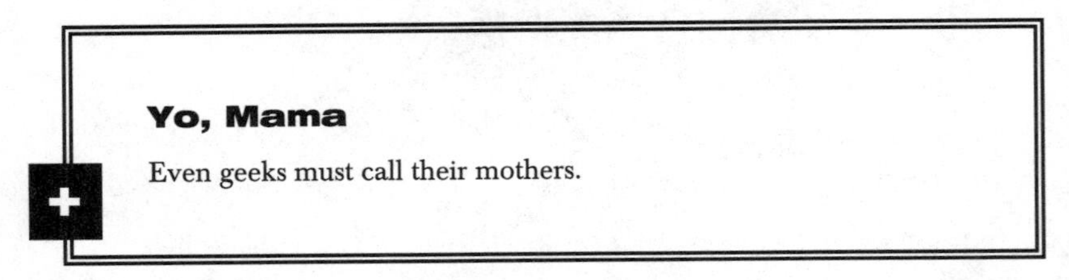

Yo, Mama

Even geeks must call their mothers.

Chapter 6 will explain Skype's SkypeOut service in detail. That optional feature connects Skype and public telephone world. Every other peer-to-peer service (with a few hardcore Internet phone exceptions) also offers this option.

You will see a variety of names used to describe this service. SkypeOut is nicely descriptive and gives the idea succinctly. "PSTN to VoIP Gateways" clangs and adds another reason to never let engineers do marketing. Yes, you now know that clunky slogan means Public Switched Telephone Network to Voice over Internet Protocol Gateway, but that's not nearly as cute as SkypeOut. Besides, it should be called VoIP to PSTN, since these gateways support calls out from peer-to-peer telephone services over the Internet to the public phone network.

Many peer-to-peer companies have configured outbound gateways for users to reach the public telephone network. Most of these cost money because they require hardware servers and lines from local phone companies connected by those servers to the peer-to-peer network. Since they have to charge money to pay for such service, they have to bill you when you use that service. And billing costs more money, and then they have to track when you make a call that must use a long distance link away from the local phone exchange hosting the gateway and on and on. You can see how this gets complicated.

Luckily, you don't have to worry about much of this at all. When you use a peer-to-peer service and call only those in your service's address book, there's never a charge. When you call out to the "real" world, there may or may not be a charge. Many cooperative groups have collaborated to shuttle Internet calls to a public telephone network central office serving the telephone number you dial. And when you do have to pay, it's pennies per minute, and only when you use it.

Virtual Numbers

Calling out from a peer-to-peer network to the public telephone network is one thing. But if your peer-to-peer service lists you as James-Gaskin inside the Skype user directory (well, yours doesn't, but mine does), how do outsiders dial that? How can they call you?

Enter virtual numbers and the inbound side of the public phone to peer-to-peer network gateway. *Virtual numbers* are traditional telephone numbers that link the public telephone network into the broadband phone world. Vonage and their competitors with analog telephone adapters that plug into your broadband modem are providing virtual numbers as well, but they don't really say that. The phone-centric companies assume you will tie a telephone number to an actual phone, just like the traditional telephone companies have always done. Even your cell phone has a virtual number if you look at it that way.

The beauty of virtual numbers for Internet Telephony is that clients are no longer bound to the telephone company central offices that, since area codes were implemented in October 1947, defined the area code prefix. Distance from the central office is critical. It sets your exchange, and therefore your area code, which is based on the area's upstream central office with the huge telephone switch. Across the Internet, a mile or a thousand miles makes almost no difference in the voice quality of the call. That means broadband phone companies can connect their network to the telephone network anywhere with any area code, and connect that number to a phone a continent away.

You live in Cleveland and want all your friends in Dallas to be able to call you with a local call? Your provider can give you a phone number with a Dallas area code, such as 214 or 972. Your friends call you on the 214 number, it travels across the Internet, and your phone in Cleveland rings. Your calls out to Dallas will be local calls, but if you call for a pizza delivery one snowy night, that call will be long distance. Remember, the area code for your Cleveland phone comes from a Dallas central office.

However, businesses love this. If you have a business in Dallas and want the people in Fort Worth (30 miles away) to think you're local, get a virtual number in the Fort Worth area code (817).

Travelers love virtual numbers as well. With softphone software on a laptop or PDA, you have your phone number anywhere you can get a broadband connection. Want to be that annoying person in Starbucks hogging a table with their laptop? Now you can double the annoyance factor by using your laptop's softphone to talk noisily with your friends back in Cleveland. Sure you could do that with your cell phone, but that costs a lot more than a softphone. Plus, everyone hates noisy cell phone users, but softphone users are rare enough to get a pass for the cool factor.

What's Old Is New Again: Internet Telephony from Phone Companies

You didn't think the traditional telephone companies were going to shuffle quietly off to retirement, did you? They have this strange idea that because they developed the world's best telephone network in the first place, they should be able to keep our business no matter the technology used to carry the conversation from phone to phone.

Acronym Alert:

ILEC = Incumbent Local Exchange Carrier and RBOC = Regional Bell Operating Company

When AT&T split, the local telephone service in the United States was split into seven companies called Regional Bell Operating Companies. These seven merged down to four: SBC, Verizon, BellSouth, and Qwest. One of these four companies controls the copper pair of wires providing telephone service to your home.

Surprisingly, these old guard traditional telephone companies aren't thrilled to see various broadband phone companies stealing all their customers. Worse, all their dirty tricks over the years (filing lawsuits to keep companies from tying into the telephone network and refusing service so often that the courts regularly had to step in and force these phone companies to follow the laws) have used up most of the good will people had for them. Remember back when most of us hated the cable company but loved–okay, *trusted*–the phone company? Now most of us hate both, or worse, ignore both and rely on our Internet Service Provider.

That said, the traditional phone companies are willing to let bygones be bygones and provide Internet Telephony if that's what we want. And they have considerable advantages they can leverage against the newer broadband phone companies.

Leveraging Copper

What's the biggest advantage the traditional telephone companies have in their favor? They own the copper wire that reaches into every home and business. Oh, and inertia is the strongest physical force, so they have that on their side, too, because it will take effort to convince phone users to switch. This will make it easy for the traditional telephone companies to offer broadband plans to save money and keep the majority of their customer base. Not only do the incumbents own the copper wires from your home, they own the central offices to which they connect. If they don't take advantage of these two giant leaps ahead of their competition, they deserve to go bankrupt.

Part of the reason the traditional telephone companies are getting into this market is simple: if people want broadband phones, and the telephone companies don't have broadband phones, customers will leave. That hurts the feelings (and bottom line) of even the biggest company.

It will be interesting to see how these companies will market these services. How do you say that traditional telephone service is old-fashioned and useless today, when you're busy signing up the majority of new customers to the traditional telephone service? How do you price broadband phones, with all the features that come standard, while charging ten dollars per feature to your traditional phone customers?

Maybe these companies have all this figured out. After all, Ford sells sporty cars, minivans, and pick-up trucks from the same dealership.

Verizon and AT&T

Of the incumbent local carriers, Verizon is the only one actively selling a deployed broadband phone service (as of early 2005). AT&T pushes their service, but since they no longer have an installed base of local telephone customers, I hold up Verizon as the example for the other three Ma Bell children.

The Verizon broadband phone service is called VoiceWing. They don't explain what the term VoiceWing has to do with Internet Telephony, however. It does have a nice feeling of freedom, though, don't you think? Then don't read the fine print about Verizon charging you in advance for each month's service, because that doesn't sound freeing.

Verizon's feature set is about average. (A description of all the broadband features available awaits you in the next chapter.) They do not have support for 911 calls as of this writing. They also offer a discount on bundled services that include a DSL line. This will supposedly stop customers already buying cable service from Time Warner from signing up with their broadband phone service.

AT&T offers almost exactly the same services at the same price point as Verizon. AT&T also offers DSL connections for Internet access, which of course is necessary for broadband phones (the first word is broadband, right?). They also work with a cable broadband modem, since the telephone adapter they ship to you works with either.

911 service comes with AT&T's CallVantage plan, although you must register your phone details and physical address with local emergency services.

SBC and Qwest and BellSouth

The rest of the incumbent telephone companies are dipping their toes into the Internet Telephony water, but haven't jumped in fully yet. Why? I believe because the cable companies are doing a better job replacing the phone for people than the phone lines are doing replacing cable. You can get telephone service over your cable TV and cable Internet connection now, but you can't get movies over your telephone wires.

Von...Van...What?

If Call*Vantage* sounds much like the Vonage name, Vonage noticed that as well. In fact, Vonage sued AT&T in March 2004, claiming the CallVantage name comes confusingly close to the Vonage name. Vonage, CallVantage, Vonage, CallVantage, hmm. No doubt the courts will have to get involved. But if you say Vonage like my friend in London, "VoNaaahhgge," there's no confusion.

There are two types of cables running to most homes in the United States: telephone and cable TV. Early on, the two monopolies coexisted warily if not peaceably. Both industries wanted to get more of the consumer's money, but they didn't know how.

Cable companies started delivering broadband Internet access before the telephone companies got DSL regulated, supported, and priced for consumers (the phone companies really had broadband years earlier but limited it to high-priced business networks). Cable companies engineered their systems to deliver a fair amount of bandwidth to the customer, necessary for delivering television, where each channel needs about 6 Mbps throughput. Cable companies have been fairly generous with bandwidth for their Internet access customers, providing 2 Mbps from early on and often upping that to 3 Mbps or 6 Mbps for no additional cost as they built up their capacity. Physically, broadband cable could deliver 100 Mbps and more to each home with backend upgrades on their facilities.

It's a tough job to sell a customer on Internet telephony over a DSL line when the customer already has a high-bandwidth cable Internet pipe spewing bits all over the place. Some 100 million or so American residences are within the service area for cable Internet service.

So SBC and Qwest and BellSouth are playing with pilot projects and selling Internet Telephony to their business customers buying high-bandwidth data-networking links from them. Cable may have bandwidth to the consumer, but the telephone companies have been supplying bandwidth to businesses long before DSL was available. So the phone companies have a large customer base of data clients eager to switch to Internet Telephony to lower the cost of their data network connections. Of course, the phone companies lose money from long distance charges the businesses avoid by using Internet Telephony, but half a loaf of customer phone payments are better than no loaf.

Dump Which?

Cable TV companies have an easier time convincing customers to dump their telephone lines than phone companies have convincing customers to dump their cable.

Remember how the telephone companies have a huge installed base of copper wires to homes? In many cases, those wires can be souped up with new flavors of high bandwidth DSL and provide enough bandwidth to even sell television services over that link (theoretically, but the few pilot projects underway here and there haven't expanded).

What if those copper wires could be replaced, or at least augmented, with a fiber optic cable? That would mean hundreds of megabits per second of bandwidth to homes, delivered by the phone companies leveraging their huge number of wire installation employees and expertise.

This is called Fiber to the Premises (the clumsy acronym is FTTP), and it has become the mythical promise of every phone company around the country when a grand plan is needed. When municipalities get upset with the slow pace of broadband access or modern services delivered by telephone companies, the phone company executives pull out their Fiber to the Premises dog and pony show. Every phone company has one small pilot project underway, usually sending high speed fiber to every home in expensive new subdivisions. They use these projects to "prove" they are working toward a better bandwidth tomorrow.

The problem for telephone companies? Customers they convince to buy broadband phones are already their customers, so they don't gain customers. Of course, keeping customers, even at a slightly lower revenue, is better than losing customers.

Redial

Many broadband phone companies focus their attention on the telephone, a device their clients are familiar with and want to keep. These phone-centric companies, led by Vonage, provide adapters to connect existing analog telephones to broadband modems.

Other broadband phone companies follow a computer-centric focus, leveraging the powerful peer-to-peer networking model so popular on the Internet. They support telephones, but only new telephones that plug into computers. Existing telephones are replaced by headsets and software phones running on computers.

The incumbent telephone companies are getting into the broadband phone business, but lag far behind the cable TV companies. Every telephone company has deployed or announced Internet Telephony, as has just about every cable TV franchise and many Internet Service Providers.

No matter which option you choose, you'll have phone number flexibility thanks to virtual numbers. And broadband phone companies no longer charge extra for long distance calls within the country, so that's more money saved by switching from your traditional telephone to a broadband phone.

FREE INTERNET PHONE FEATURES THAT YOU'RE PAYING FOR NOW

Telephones used to be simple. Dial. Talk. Hang up. Of course, a few decades ago our choices were simple but unpalatable. Pick up the phone you leased from AT&T. Dial (physically, with your fingers going in circles around the actual dial) a number. Talk through a decent local connection, but a poor long distance connection. Pay dollars per minute long distance prices. If you didn't like that situation, you could, ah, write letters, because there was only one phone company.

Movie Alert

Learn all about TPC (The Phone Company) in the marvelous 1967 paranoid comedy thriller starring James Coburn called *The President's Analyst.* Great movie.

Today all phones do more than just dial and talk. Everyone has become spoiled by the modern features every phone has, such as automatic redial, one-button dialing, and a variety of other amazing phone tricks. And the services from the telephone company, such as Caller ID, call forwarding, call waiting, *69 return call, call blocking, and voicemail, among others, thrill many people all but one day per month.

That one day, of course, is when they pay their phone bill. Each of those services costs money–sometimes a lot of money.

Saving Money

You can't judge how much you save until you know how much you spend for traditional telephone service. Table 3-1 shows the details from my bill at the start of 2005.

TABLE 3-1. SBC monthly statement breakdown

Service/fee description	Amount
Basic Local Service—Residence	11.05
Touchtone	.18
Total Monthly Service	11.23
Surcharges and Other Fees	
Federal Subscriber Line Charges	5.21
911 Service Fee	.75

TABLE 3-1. SBC monthly statement breakdown (continued)

Service/fee description	Amount
Federal Universal Service Fee	.90
Expanded Local Calling Service	.08
TX Rate Group Reclassification Surcharge	.93
Municipal Charge	.80
Total Surcharges and Other Fees	9.13
Taxes	
Federal (Local Charges)	.59
Federal (Non-regulated and Toll Charges)	.00
State and Local (Local Charges)	1.61
State and Local (Non-regulated & Toll Charges)	.01
Total Taxes	2.21
Total Plans and Services	22.57
SBC Long Distance	1.21
Total Current Charges	23.78

Take another look at the first figure in Table 3-1: Basic Local Service–Residence. That's the number the phone companies like to quote. Notice the amount: $11.05. Darn cheap, considering all that you get for it, isn't it? A world-wide telephone network at your beck and call, and it sits waiting in your house (okay, my house) for $11.05 a month.

But look at the bottom number, the Total Current Charges: $23.78. This number is more than double the "basic" service, even if you add the Touchtone fee in there. That's serious fee inflation.

Fee Bull

Touchtone fee? They charge me for Touchtone service still, after all these years? I bet if I dug up an old dial phone, they'd charge me even more for "retro dial" service.

Notice the last little tag, the SBC Long Distance? Guess what–I didn't make a single long distance call on this line. They're charging me for the possibility I might make a long distance call.

Okay, say the Bell apologists, but $23.78 is still less money than Vonage charges you, which is $24.99 per month.

Wrong.

Vonage (and competitors offer something similar) has a Basic 500 plan that includes 500 minutes of phone calls to any phone in the United States and Canada. In other words, you can get a Vonage phone line, talk 500 minutes to the farthest point from you in the United States or Canada, and pay no long distance charges. You'll pay $15, versus the Bell charge of $23.78 per month that does not include a single minute of long distance calls.

To be fair, there is no time limit on my SBC local calls. I can talk 24 hours per day locally for the same $11.05–er, $23.78. With the $15 Vonage plan, I get the first 500 minutes free, but must pay 3.9 cents per minute on every call afterwards.

Of course, the Vonage call can be from Miami to Anchorage for the same 3.9 cents per minute after the first 500 minutes, which is cheaper than any long distance rates from SBC (by half at least, and often more depending on the called location). And if you get the Vonage Premium Unlimited Plan, you can call every minute of every day for $24.99. Now that's more than SBC's charging me per month (by $1.21) but it also includes all the long distance (U.S. and Canada) calls I can stand to make. Do you consider unlimited long distance calls a great deal for $1.21 per month? I do.

Notice what my phone bill is missing: any extras whatsoever. Few people bypass Caller ID and Call Waiting and the like, and I'm about the only one in the world with a "naked" residence phone line (according to my teenagers). But SBC doesn't want us drilling down into these extras charges for reasons that will become obvious in the next section.

SBC Charges Versus Broadband Phone Charges

Internet telephone companies can add features much more easily, and therefore much more economically, than the traditional telephone companies. In fact, the $14.99 or $24.99 you spend with Vonage (or competitors like VoicePulse and BroadVoice) includes features worth $61.15, according to SBC and other traditional telephone companies.

Here are the features from the three aforementioned broadband phone companies:

Vonage features	VoicePulse features	BroadVoice features
Voicemail Plus	Enhanced Voicemail	Anonymous Call Rejection
Caller ID with Name	Online Account Center	Do Not Disturb
Call Waiting	Phone Shortcuts	Call Forwarding Always
Call Forwarding	7-Digit Dialing	Flash Call Hold
3-Way Calling	Contact Lists	Call Forwarding Busy
In-Network Calls	Distinctive Ring	Flash Call Transfer
Traveling with Vonage	Speed Dial	Call Forwarding No Answer
Area Code Selection	Anonymous Call Block	Flash Call Waiting
Call Transfer	CallerID Block	Call Notify
Click-2-Call	Directory Assistance Block	Flash Three-Way Call
Call Return	Do Not Disturb	Call Return
Caller ID Block	Filters	Last Number Redial
Repeat Dialing	International Call Block	Call Waiting
International Call Block	Telemarketer Block	Speed Dial 8 & Speed Dial 100
Ring Lists	Call Forward	Caller ID - Name Retrieval
Call Hunt	Line Unavailable Forward	Voice Messaging/Management

The cost of the features for each of these services, beyond the monthly fee: $0.00 (zero, zip, nada).

Here is the list of features from SBC, and the price per month for each:

Feature	Price
CallNotes Plus (voicemail)	$9.95
Caller ID	$9.95
Call Waiting	$2.80
Call Waiting ID	$5.00
Personalized Ring	$2.95
Call Forwarding	$5.00
Selective Call Forwarding	$1.50
Auto Redial	$4.00
Call Return	$5.00

Feature	Price
Speed Calling 8 (one-button shortcuts for 8 numbers)	$5.00
Call Blocker	$5.00
Conference Calls (3 way)	$5.00
Total	$61.15

Ouch.

Many of the popular features from the traditional telephone companies are bundled together, usually with some price break (although sometimes the reduction is pennies). Some users appreciate the savings, while others resent being forced to pay for features they don't want in order to get a good price on the features they do want.

Notice there are some features available from broadband phones that traditional phone companies can't match, such as virtual numbers. Notice also that the $61.15 from SBC includes no long distance minutes, while each of the broadband phone calls include all the calls you make within the U.S. and Canada as part of the basic monthly fee.

Calls at Zero Cents Per Minute

Since all major broadband phone services include long distance calls within the U.S. and Canada as part of your monthly fee, there are no long distance charges unless you start dialing numbers beyond the borders of the U.S. and Canada.

If you plan to spend plenty of time chatting overseas, you have two choices: check out the International Plans offered by broadband phone companies, or look to Skype and other computer-centric options. Sometimes the broadband phone services offer a flat rate plan for international calls to a set number of countries, and sometimes they offer per-minute pricing. The rates are different for each country under the per-minute plans, and they sometimes change without warning.

The computer-centric plans never charge a penny for either their service or international calls when connecting from computer to computer. This means both the caller (you) and the callee (the person you want to speak with) must be connected to a computer and online to make a call. But when even the best international prices can be close to a dollar per minute (or many dollars per minute when a bulky satellite phone set is involved), sticking close to your computer may save you a fortune.

Traditional phone companies charge much more per minute for international calls than do the broadband phone companies. So if you make a lot of international calls, you have every reason in the world (and your wallet) to switch to a broadband phone service immediately.

Handling, or Not, Calls

If you are a slave to the ringing phone, you may not know the bone-deep joy of ignoring calls when you find them inconvenient. I recommend you hang a mental "Do Not Disturb" sign up sometime, let the answering machine earn its keep, and enjoy not answering the phone for a change.

However, even when you want to answer the phone, you may not be able to do so. Many features from broadband phone services allow you to control, rather than react to, incoming phone calls. The following sections describe these. I've also included a list of prices from various traditional phone companies as well as from broadband VoIP providers.

Caller ID

- SBC: $9.95
- Verizon: $7.95
- BellSouth: $9.00
- Qwest: $6.95
- Broadband phone services: $0.00

Want to know who's calling? The traditional phone companies count on that desire, to the tune of $7 to $10 per month per line. Qwest even has the, ahem, courage to charge for installation, but it could be they're the only ones listing it on their web site where I can find it.

Getting Caller ID is a double whammy for some people: they pay for the service, and they need to buy new phones to actually use the service. The traditional telephone companies still sell little Caller ID boxes that sit beside the phone for those users with older phones that don't have Caller ID support.

Want to know the sad part about these Caller ID charges? All phones calls now carry the information needed to display the caller information. There is no longer any extra expense borne by the traditional telephone companies to rationalize their high Caller ID prices.

Call Waiting

- SBC: $2.80
- Verizon: $0
- BellSouth: $7.50
- Qwest: $5.50
- Broadband phone services: $0.00

Can't stand to miss a call, even when you're on one? Call waiting service has been available from traditional telephone companies for quite a while, and the general public seems to like it.

So, of course, the broadband phone services include call waiting. After all, when you're doing all the phone call handling via a computer rather than a telephone switch, such features are simple to add.

Call Forwarding

- SBC: $5.00–$7.50
- Verizon: $1.80–$2.50
- BellSouth: $5.00–$15.50
- Qwest: $3.00–$3.50
- Broadband phone services: $0.00

When someone calls you where you aren't, isn't it nice to have the phone ring where you are? That's Call Forwarding, and the traditional telephone companies made a big deal about this "advanced" feature when they made it available.

There are multiple ways to handle call forwarding, however, such as:

- Call Forwarding Always, Busy, and No Answer
- Call Forwarding Selective (certain numbers are forwarded)
- Call Forwarding–ring at multiple phones at once

You may also see these service listed as something like "Multipath" or "Simul-Ring" when a call can go to more than one phone at a time. Remote access to control call forwarding comes with a steep price at BellSouth, but once again, it's free at the broadband phone service companies.

Note that if you forward calls to a cell phone and have both ring, but the cell phone voicemail picks up sooner than your desktop phone, the message will never make it to your desktop phone.

Conference Calling

- SBC: $5.00 (three-way only)
- Verizon: $4.00 (three-way only)
- BellSouth: $6.00 (three-way only)
- Qwest: $3.50 (three-way only)
- Broadband phone services: $0.00

This is a favorite teenage feature, judging by my daughter and her friends. Their cell phones can do it, so they expect their desk phones to let them add another caller to the phone party whenever they want.

The traditional telephone companies and phone-centric broadband phone companies are all limited to three-way calling. Skype, however, offers up to five callers in one conference (they are talking about increasing that number). You can't get much real work done that way, but it's a nice feature for the marketing department to tout, don't you think?

Other Features

People love features, or at least some people love as many features as they can get, while others really love a few features and don't care about the rest. Some care about no features, until they find one that directly solves a problem they have. Then they too become feature-mongers.

There are not many more features available from the traditional telephone companies that appeal to very many users. Prices vary widely between traditional telephone companies, as you can tell from the prices listed already. When you do find a service you like, you have to pay for it month after month after month.

Broadband phone services have a long list of features still to go. These next features discussed are available only from broadband phone services. The traditional telephone companies can't match the new technology with their old equipment.

Choose Your Area Code

This is one of the coolest advantages of broadband phones, and one that the traditional telephone companies absolutely cannot copy. Even if broadband phones cost the same or more than traditional phones (they don't, but imagine they do for this example), the ability to choose the area code for your number would make them mandatory for many people.

The nomenclature has yet to clarify, so broadband phone companies use different terms for the same service. "Virtual Numbers" means something completely different, at least in certain circles, and appears later in this chapter in the "Optional ($$) Features" section.

The great thing about this is the ability to pick your phone number from any area code served by the broadband phone company, rather than being limited to a specific geographic area as forced by the traditional telephone companies. You moved and want a phone number that's a local call from all your old friends and family back home? You got it.

Vonage calls this ability to live in Cleveland and choose a phone number that's local to San Francisco *Area Code Selection.* Sounds good. VoicePulse calls this Available Area Codes. BroadVoice calls it Number Availability, but don't treat any of these as gospel because marketing vice presidents will change them now and then.

The important part to remember is that your broadband phone company can provide you a phone number from anywhere in the U.S. where they have connections to the local telephone company. This is great for businesses, who want to give customers a local number to call, knowing customers won't dial long distance. It also makes the business appear local.

Limitations exist, depending on how much support your broadband phone company has throughout the United States. None of the broadband phone companies offer every U.S. area code, or even local numbers for all major cities, as of early 2005 (but that will change soon). And if you check the service areas of the various broadband phone companies, I promise you will find the cities you need are covered by one of these carriers.

Since area codes cover multiple cities (especially suburban towns), you may find your exact city doesn't have a number available but another town nearby does. Why? The traditional telephone company's central office in your exact town may not yet support your broadband phone company, or the block of numbers allocated for the broadband phone customers may all be used. This doesn't matter, because you'll get the area code you want, or at least an area code that is a local call for the area.

Cold Pizza

If you call your local pizza parlor for delivery on your broadband phone, make sure they have the right city, or the pizza will be cold and the delivery charge will be horrendous when it arrives (assuming your pizza place tracks you by your phone number as mine does).

Be careful–if you have a broadband phone with a distant area code, you probably need a second phone, or at least another number for your broadband phone. Otherwise, if your neighbor calls you, it will be a long distance call for them.

Look at the map from BroadVoice in Figure 3-1, which shows the screen picking an area code. There are three funny things about this map, which I'll get to shortly.

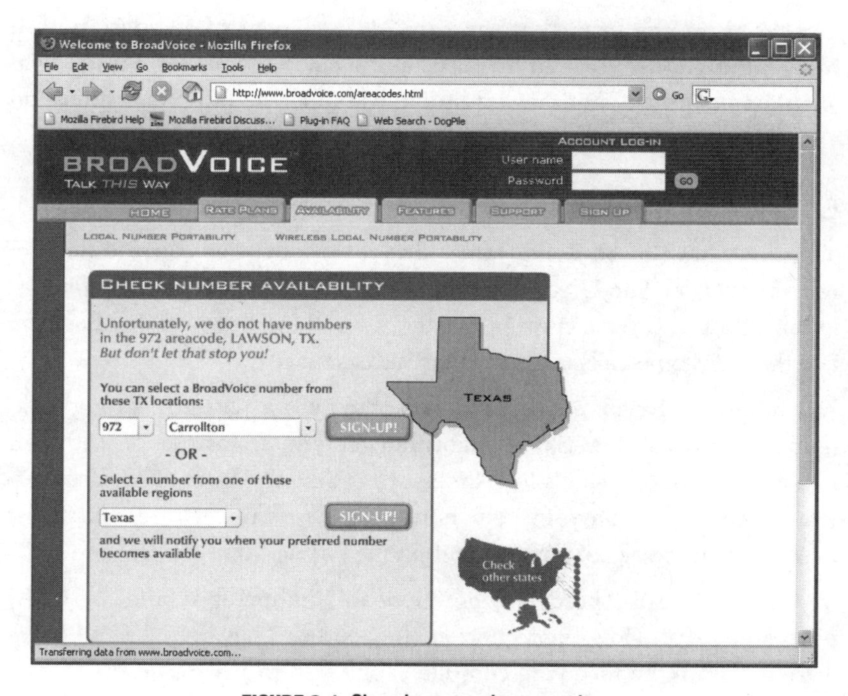

FIGURE 3-1. Choosing your phone number

I'm in a suburb of Dallas called Mesquite (doesn't that sound like Texas?). My area code is 972, used in the Dallas area when the 214 area code started running out of numbers. 972 numbers are the donut around the Dallas 214 phone numbers.

My suburb's traditional telephone company's central office doesn't have any available numbers, but the central office in Carrollton does. If Dallas is a clock, then Mesquite is at three o'clock, and Carrollton is at eleven o'clock and about 45 miles away, but they're both local numbers in the Dallas area.

Okay, so what's funny? First, BroadVoice doesn't have numbers in Mesquite, but they still have some in Dallas with the 214 area code. Second, my local phone number can terminate at a central office 45 miles away, but it's all local for the broadband phone companies. Finally, my central office still carries the town name of Lawson, although Mesquite is where the central office is and where I am. I'm not even sure Lawson is still incorporated. Told you the traditional telephone companies move slowly.

You've heard the term "cut the cord" in a variety of settings. With a broadband phone, you can literally cut the cord of your local traditional telephone company and have a little piece of home (a phone) sitting on your desk far, far away.

Encrypted Conversations

You don't have to be Bond, James Bond, to want your conversations to remain private. The U.S. government doesn't like encrypted telephones in the hands of private citizens, but there's nothing they can do about the fact that all Skype calls are encrypted automatically.

Not only does Skype encrypt the calls, they use a seriously strong technology to do so. The Skype algorithm (computer instruction set) uses 256-bit encryption. That's four times better than the level of wireless network security you have in your home. That's more secure than almost every nonmilitary U.S. government phone encryption service.

Skype relies on the hardware power of the computer system (or PDA) processor to provide encryption at the source. They rely on the hardware power of the receiving computer (or PDA) processor on the receiving end to decrypt the voice stream and turn it back into sounds.

Decrypt 'Em Danno

If you're a law enforcement official and want to eavesdrop on conversations, you hate Skype.

The phone-centric companies like Vonage do not encrypt their voice streams. Why not? Two reasons. First of all, what they don't admit is that adding encryption technology to the broadband routers used would be difficult and expensive. Skype's software product makes them much more flexible. Second, the phone-centric companies rely on the traditional telephone company networks to link to non-broadband phones. Federal law already demands the phone companies allow law enforcement access to tap into the telephone calls of citizens (at least they're supposed to get a court order before eavesdropping, but sometimes little details get overlooked). So the phone-centric broadband phone companies can't guarantee encryption from end to end.

Skype software on both ends of the connection can guarantee control over the entire length of the call. Using the SkypeOut service to call to non-Skype phones kills that idea, however, leaving those conversations unencrypted over the traditional telephone lines.

Free Branch Office Connections

Have a small business? Want one? Have a partner? Need to talk to your partner in another city? Broadband phones will save you a fortune.

One enterprise telephone study I saw in mid-2004 said 90% of long distance charges for some huge companies are employees talking to employees. Isn't that what a water cooler is for, or is that too old-fashioned?

Your broadband phone works great as a local phone number, and with all the business plans and most of the residential plans, long distance calls within the U.S. and Canada are free. Have a partner in Chicago and you're in Cleveland? Call each other for free.

Companies buying a new business phone system based on Internet Telephony will receive a huge surprise: a magic extension capability. With an intelligent Internet Telephony phone, the link between the phone controller–which used to be a PBX (Private Branch eXchange) but now it's a computer–can include the Internet. You know how you pick up a desk phone at work, dial an extension, and talk to your buddy a few cubicles down? With an Internet Telephony business phone system, that cubicle could be in Cleveland while you sit in Chattanooga.

Domestic long distance bills will become history when you get a broadband phone system. International calls are a different story, but standard long distance, between one place in the U.S. or Canada and another place within the U.S. and Canada, will become a not-so-fond memory for you. Switch to a broadband phone and talk cheap.

Optional ($$) Features

Not every service from every broadband phone company comes wrapped inside the low monthly price. Some services, especially those which require a considerable amount of new technology to power or have no comparable service in the traditional telephone world, require a few more dollars per month.

You shouldn't find this surprising or out of place. After all, if SBC can take an $11.05 fee for local phone service and jack it up 50 more dollars for their services (admittedly some of those dollars are federal and local taxes), you can pull another few dollars out of your pocket for some amazing new services not available anywhere else.

Virtual Numbers

Although you can choose any area code you want with your broadband phone service, one number from a distant area code may not be enough. And when you use a broadband phone service as your only telephone, as hundreds of thousands do already, you may want to add remote area code support.

A *virtual number* is one from a different area code that rings at your local phone number on your broadband phone. Similar to a forwarded call, these numbers are used for incoming calls only (you can already call the remote area code for free as part of your no-charge long distance calls) and are great for family and far-away friends.

Most major broadband phone services have virtual numbers, and the fees range from $5 to $10 per month. On the remote end, someone dials your virtual number just like any other number in their local area (no long distance call needed). On your end, the phone rings and it's someone from a long way away. Chat and enjoy.

Set up your virtual numbers (you can have as many as you want) through your web account with your broadband phone provider. A few mouse clicks, and Aunt Harriet in Hoboken can call you in Cleveland for free in no time.

Businesses love these virtual numbers. Want a presence in multiple cities but want to answer only one telephone? This is your solution.

800 Numbers

Look at this as a Super Size virtual number, and you'll get plenty of enjoyment without extra calories. And don't think this is only for businesses with big budgets, because this service is now used by all sorts of normal (maybe slightly special) people, including Aunt Harriet in Hoboken.

Why get an 800 number rather than a standard number? What if you and your spouse come from different hometowns, and both want people to be able to call for free? What if your family is scattered far and wide and one area code doesn't come close to handling them all? What if you have children in college, and they need to call home (for money, of course)? The 800 number, or virtual toll-free number, makes all this possible.

Here's a quick scan of pricing from three major broadband phone services:

Vonage

$4.99 per month, 100 minutes free, 4.9 cents per minute thereafter

Packet8

$4.95 per month, 100 minutes free, 3.9 cents per minute thereafter

BroadVoice

$1.95 per month, 60 minutes free, 2 cents per minute thereafter

The virtual toll-free number doesn't cost any more than any other virtual number (most services offer those at around $5 per month, but you're not charged extra for the minutes). A certain amount of minutes are free with each of the toll-free virtual numbers, and thereafter you pay a few pennies per minute. There are also installation fees for some services, so check that so you won't be surprised.

If you or someone in your family travels regularly, this toll-free virtual number will pay for itself in only one trip a month. Have you priced phone calls from hotels lately? Even Al Capone would be embarrassed to charge as much *vigorish* on fellow mobsters as the hotels brazenly charge you today.

Think a cell phone is a better answer than a virtual toll-free number when you want to call back home from the road? Pay a couple months of roaming charges and see if you don't like $5 per month better. I can almost guarantee that the cell phone will cost many times more than the monthly fee for an extra number on your broadband phone service account.

Video Phones

The World's Fair in New York in 1964 made news for many reasons, and one of the most famous was a demonstration of a video phone. We have now been waiting 40 years for the traditional telephone companies to deliver this, and we have nothing. You can buy a $10 camera for your computer that does a better job than any video phone you can get from a traditional telephone company today. Once again, the geniuses behind Internet Telephony win the innovation race.

Okay, I can't tell you the bugs are worked out and video phone service is fantastic today. But it's getting better, and your broadband phone service of choice either has an option now, or will sometime soon.

One company pushing their broadband video phone service is Packet8 (*http://packet8.net*). Figure 3-2 shows their web page.

I like "Speak in Color!" as a tag. It doesn't make any sense, literally, but I understand exactly.

Since video connections require more bandwidth and other considerations, I'm saving the in-depth discussion for Chapter *8*. Besides, that's a nice connection to the Packet*8* image in Figure 3-2, isn't it?

Redial

Some people use basic phones and nothing much else. They're still paying more for a traditional telephone line than they will pay for a broadband phone.

Some people love every fancy phone feature they can get. They will cut their phone bill by half or better when they switch to a broadband phone.

Don't feel sorry for the traditional phone companies and how bad their feature list with huge monthly fees looks compared to broadband phone services. Ma Bell had her chance to innovate and drop prices for well

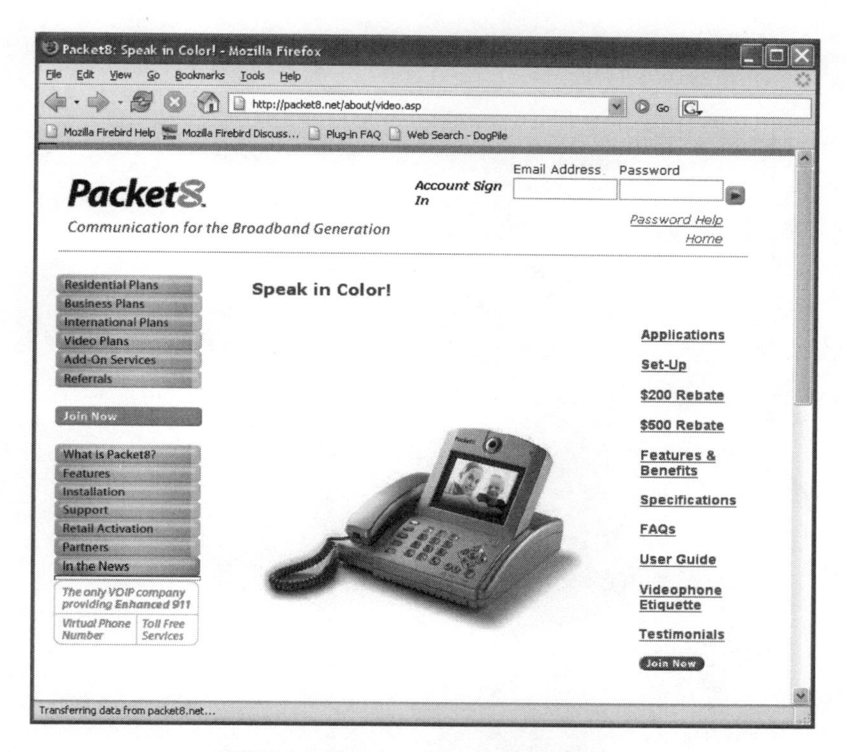

FIGURE 3-2. All-in-one video phone and service

over 100 years, and she didn't do it. In fact, the traditional telephone companies did such a good job depressing innovation, we're all amazed at the features available from broadband phone companies.

We should have gotten these features years ago. The traditional telephone companies didn't deliver, and they still can't deliver. If you want features, you need a broadband phone, period.

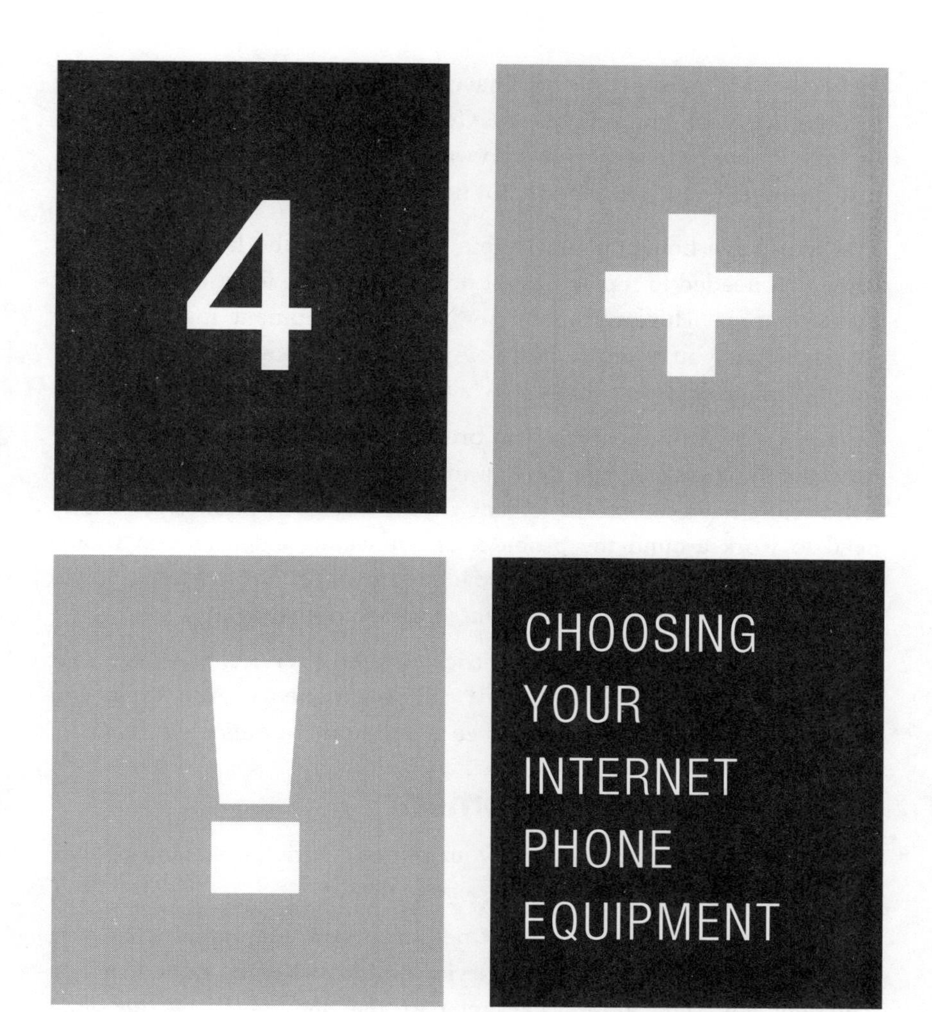
4
CHOOSING
YOUR
INTERNET
PHONE
EQUIPMENT

If you're like most people, you haven't thought about telephone equipment for your home for years. Cell phones you worry about, especially if you have teenagers who want the newest camera phones with text messaging and sexy covers, but not your home phone.

You may have bought a new home telephone in the last decade only when you needed to replace a unit dropped into the toilet or chewed by a new puppy. Thinking about new phone equipment may be more stressful than you want, honestly. After all, you bought a phone that shows the Caller ID information, what else do you need?

In most cases, with most types of broadband phone services, you don't need anything else. If you don't want to think about new phones, you don't really have to, except for one detail: extension phones. You may need to work around the problem of connecting extensions in rooms where you don't have network broadband connections available, so I'll talk about the new phone equipment you should investigate.

Those interested in computer-centric telephone service, however, will need to buy a headset for the best calling experience. And when you want a video phone, you'll need to go to the store (or online) yet again.

Phone-Centric Equipment

Where do you get equipment for your phone-centric broadband phone? It comes in the box with your new service details.

The easy answer is that you don't need any new equipment when you sign up for a broadband phone service besides what the phone service will send you. The most important part–the phone–will be the same phone you use now unless you upgrade it.

Vonage, for example, includes a new broadband router with telephone connections as part of your service. When you buy one of these routers and service plans at a store locally (RadioShack, Circuit City, Best Buy, etc.), service details from Vonage or one of the other broadband phone companies come in the box with the router. Figure 4-1 shows a close-up by Best Buy for a Linksys router that includes Vonage phone service.

That said, your service may not include all the components, or you may want to integrate a new broadband phone system into your existing home network. In that case, you may need some new equipment.

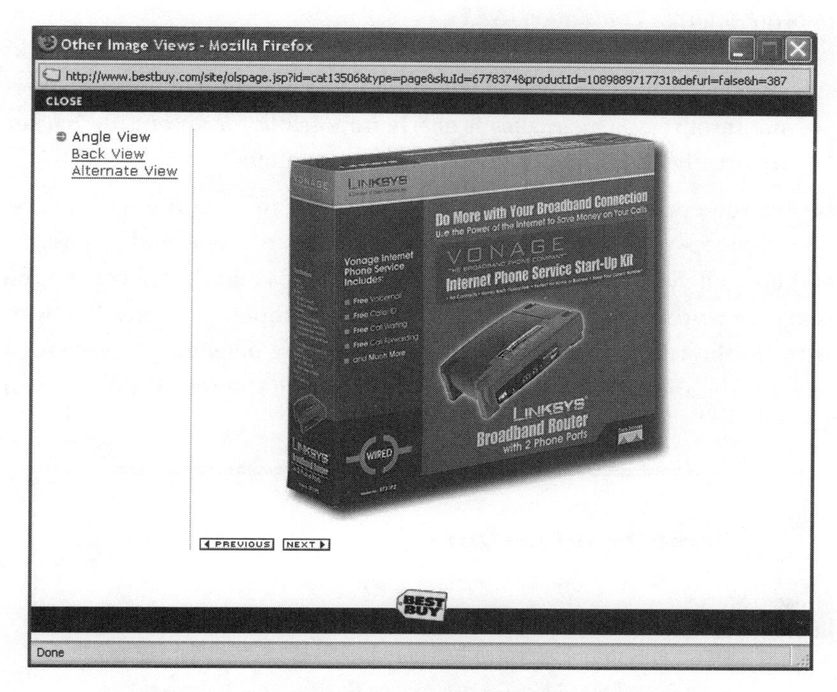

FIGURE 4-1. One of the many retail outlets partnering with Vonage

Get a New Broadband Router (Free)

If you don't have a home network already, or you want to upgrade to a new router as a "free gift" from your new Internet telephone company, you will get a router customized with telephone plugs added. They look and perform just like any other broadband router, but they support telephones as well as computers. It will likely come with your phone number already programmed inside for easy installation.

A router sits between your broadband modem (either cable or DSL) and all your internal devices. Security becomes much easier when using a router, since the router provides these security features:

Firewall

Blocks outsiders from getting into your network by blocking incoming packets that aren't a response to outgoing packets.

Network Address Translation (NAT)

Hides the actual network addresses for all your internal network-connected devices by changing their address when they use Internet resources. This makes it nearly impossible for an outside hacker to directly attack one of your internal computers.

Newer routers with wireless networking include the security tools necessary to keep your wireless connections reasonably safe and secure. In addition, all your wireless devices (usually laptops and PDAs) receive the same network security protection as your computers connected with wires to the router, but make sure your wireless network has upgraded beyond the older, insecure WEP standard to the newer WPA or even WPA2.

Broadband Details

If you want more information about your options for broadband service, and how to network computers and other devices within your home or small business to broadband, check out my book *Broadband Bible: Desktop Edition* (Wiley) at *http://broadband.gaskin.com.*

Linksys is a company with excellent marketing presence and shelf space at major retailers like Best Buy and Circuit City. They have made deals with many of the broadband phone services to provide the routers and other necessary equipment. Although Linksys is now owned by Cisco, the leading network equipment provider for big companies, the retail space remains important and the Linksys partnerships will continue.

Take a look at the most feature-packed router with phone support, the Linksys WRT54GP2. The WRT is Linksys shorthand for Wireless Router (but it includes wired connections of course), 54G for the high speed wireless network 802.11G, and P2 for the two phone ports. A consumer product, this router uses Linksys's "friendly" case in blue and black with a single wireless antenna (see Figure 4-2).

Linksys isn't the only company making routers with broadband phone support, of course, and one of their main competitors in the home network market, D-Link, has several nice products as well. The Lingo

FIGURE 4-2. A multifeatured router with phone support from Linksys

broadband phone service makes it easy to buy a variety of third-party routers and connect them to their service, but they have a special license deal with D-Link. Figure 4-3 shows a shot of the D-Link and Lingo partnership router that shows the plugs on the back.

All these routers include the same minimum number of ports, or plugs, on the box. First, they have to have power. These use a large wall wart (external power supply) with a small round connector on the router to keep the size of the router small. You can see the power connector plug on the far right side of the D-Link router in Figure 4-3.

Each router needs a plug that connects to your existing broadband modem. They always label these WAN (Wide Area Network).

In a nicely complementary acronym, there are plugs for the LAN (Local Area Network). Some routers have a single LAN port, some have three (Linksys) plugs, and some have four (D-Link). D-Link labels these as "4 Auto-Sensing Ethernet Ports" in Figure 4-3.

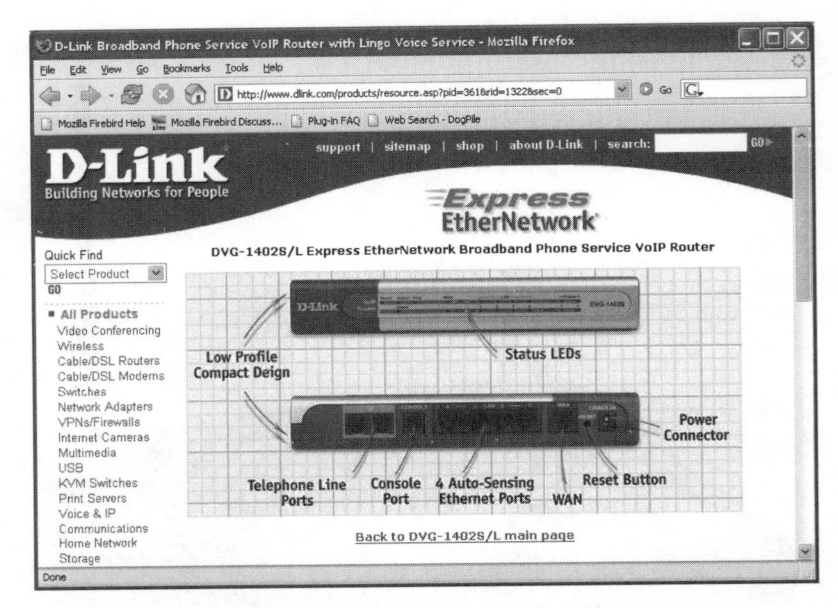

FIGURE 4-3. D-Link's version of a broadband phone router

The sign-up package you receive will include the cables necessary for connecting the WAN and power connectors, but you will need to provide your own plugs for your existing computers and other devices (game consoles, storage devices, etc.). If you already have a network, you already have all these cables.

On the far left of the D-Link image in Figure 4-3 are two telephone ports (plugs). Plug your existing telephone into one of these ports, and you're just about ready to start calling.

Routers like the Linksys and D-Link shown here are less than a hundred dollars (sometimes way less). You can get a discount on your broadband phone service from some vendors if you provide your own equipment. Look for this type of equipment at these places:

- Any large technology store with a computer department (Best Buy, Circuit City, Fry's)
- Large office supply stores (Office Max, Office Depot, Staples)
- RadioShack
- Online retailers

Acronym Alert

RJ-11 and RJ-45: RJ-11 connectors are telephone plugs, and have a maximum of four pins that support two pair of telephone wires. Some RJ-11 connectors have only two wires and therefore only two pins installed. Ethernet cables are officially called RJ-45 and have eight wires. These two types are not interchangeable. When you read a poorly written manual that doesn't explain that RJ-11 is just a phone cord connector, think of the 11 as two wires for one telephone line, leaving RJ-45 as the Ethernet network connector.

If you bought your home or small business network router more than a year ago, a new one will include better security features and easier management tools. The broadband phone service will throw in the router for free, so why not upgrade on their dime?

Use Your Current Router

Reasons not to upgrade your router on the broadband phone company's dime are if you have a new or specialized router already or need features these home-level routers don't provide. This especially applies to small businesses or home offices, particularly where your home office computer must connect to another company office and therefore you can't replace the router without getting their approval.

Acronym Alert

ATA = Analog Telephone Adapter

Analog Telephone Adapters is a term that makes little sense because just analog (normal phones like the ones you already have) phones need to be converted. Digital phones don't need an adapter because they're already digital, so they should call this a telephone adapter and skip the analog tag.

If you have an existing router you want to keep, a telephone adapter such as the one shown in Figure 4-4 will connect to your telephone on one side and your router on the other. As you might expect, there are no plugs for other network devices.

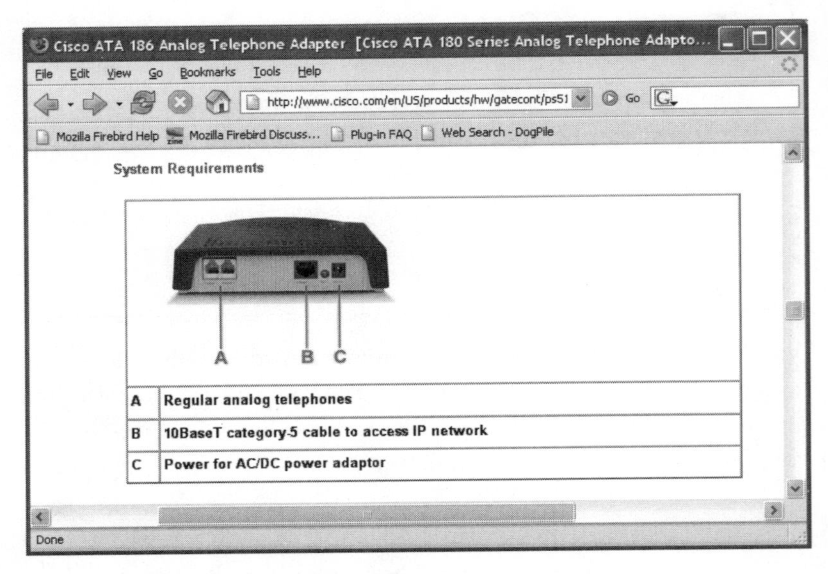

FIGURE 4-4. A widely-used telephone adapter from Cisco

Think before you go this direction, because routers with broadband phone support include features a standard router may not, especially if the router is more than a year old. For example, Quality of Service (QoS) is a range of technical features designed to ensure reliable voice conversations. Large networks such as the Internet often have little glitches and delays here and there, and the Quality of Service features help smooth those rough edges and keep the voice traffic normal-sounding. General quality for voice transmissions over the Internet is better now than ever before, but including Quality of Service support keeps the conversations sounding good.

Other features you will miss with a telephone adapter rather than a router designed for voice support are the management administration screens designed for your telephone connection. You will have some controls through the web interface at your broadband phone provider, but special routers include other features (such as status screens and easy router firmware upgrades for advanced features) that you can't get with a telephone adapter.

Notice there is a single Ethernet connection on this device. Just to be confusing, they didn't label it RJ-45 but 10BaseT, the official name for Ethernet cables made of twisted pair wiring. Your cable from your existing router will connect to plug B, and your existing telephone will connect to plug A. And as with the D-Link router, you can connect two phone lines to this single device.

Different broadband phone services will use different telephone adapters, but they will all look like this for the most part. One service even calls their adapter a *gizmo*, which I believe goes a little too far down the friendly scale and dips into the juvenile range. But you may be happy getting a gizmo for your phone, and if so, I'm happy for you. Just remember your gizmo plugs into your thingamabob.

You can buy telephone adapters at the same place you can buy broadband phone routers, but you may have trouble finding them. The marketing trend is to sell the entire router so the broadband phone service is sure you'll be getting a new router that has some extra Internet Telephony support features built in.

Adding Extension Phones

Everyone loves extension phones because we hate to jump up and start running when we hear the phone. Call us spoiled by the cell phones in our pockets, but running from room to room to answer a wrong number isn't much fun. When we get a wrong number, or worse, a telemarketer, we don't appreciate our impromptu exercise program because then we're too out of breath to curse at the telemarketer.

Unfortunately, your broadband phone must be plugged into your broadband router or telephone adapter, and you only have one of those units. Certain digital phones can be plugged directly into your Ethernet network, but you probably don't have Ethernet RJ-45 ports all over your house, either.

Answer? Go cordless. Not just so you can carry the phone around the house, but because new cordless phones include optional handsets that you can put all over the house. All the multiple handset phone systems (from multiple manufacturers) support at least four total phones, and many support eight. If you have more than eight phones in your house, your big home indicates that you have enough money to look into custom solutions or invest in a small business telephone system. The rest of us, with two or three or four total phones, will be in good shape with a new cordless system.

Motorola, a long-time maker of phone equipment as well as telephones, is a good example of a manufacturer with several models of expandable cordless phone models. Figure 4-5 shows one of their most popular models.

FIGURE 4-5. Multiple views of a multiple handset telephone

Many models include an answering machine. All include Caller ID support on the main unit and the extension handsets.

Every expandable cordless phone includes features that may surprise you, as they did me. I didn't expect the Motorola phone to act as an intercom, but it does. It also acts as a room monitor, so you can save on baby monitors if that's a current or upcoming requirement (if so, congratulations). I don't know if I can eavesdrop on my teenagers with this, but I'm going to try.

The Motorola system I'm testing did something that surprised me. Although the Motorola MD761 includes its own answering machine, I tested it with Vonage voicemail by turning the phone's answering machine off. When I had a voicemail message waiting on my Vonage phone line, a light on the Motorola base unit blinked on and off. This tells you there's a message without having to log into the Vonage web site or pick up the phone and listen for the voicemail sound. Very handy, and another reason to get the new router from your broadband phone service.

You have many options for expandable cordless phones. I did a quick check online, and here's how many units I found when looking for expandable cordless phones:

Retailer	Units
OfficeMax.com	15
OfficeDepot.com	14
Staples.com	15
BestBuy.com	24
CircuitCity.com	24

This doesn't count the number of cordless expandable telephones with support for multiple lines. And there are some phones with a regular corded handset for the base station and cordless phone extensions, for the traditionalists among you trying to slide gently into the modern phone world.

Yes, buying new telephones costs money. Some of your savings from switching to a broadband phone will be lost. However, in a few short months, the phones will be paid for and you'll have years of trouble-free broadband phone usage with full advanced phone features and even fuller savings.

Computer-Centric Equipment

Let's make this clear: if you want to use a computer-centric phone system, you need a headset or USB telephone handset. Yelling into a microphone in front of your computer while hoping the speaker echo doesn't overwhelm the conversation gets old real fast.

Headsets are by far the most common right now, but telephones built for computer-centric telephone systems are starting to appear. You can even buy one from AOL for a much better price than you can get the same phone from other sources (AOL Keyword: Broadband Gear).

Where do you get headsets and USB phones? You can get headsets at every location listed earlier in the section "Adding Extension Phones" (such as the office supply stores and other stores with decent-sized computer departments).

USB phones are more difficult to find on store shelves. In fact, I've not found any in the Dallas area yet. You'll have to go online for these unless you live in a major technical center (Silicon Valley) or a huge retail paradise (New York).

There are two ways to connect a telephone or headset to your computer: through the USB connector or through the Speaker Out and Microphone In plugs. Most desktop computers made in the last five years have the speaker and microphone plugs built onto the motherboard, and all optional sound cards have these plugs as well. Notebook computers will also have these ports. Computers made in the last six or seven years all have USB ports. The newer USB 2.0 ports offer great speed advantages, but the earlier USB ports are more than fast enough to support a voice stream.

Most headset makers have both USB and analog (speaker out and microphone in) plug options. A few makers are behind the curve on the USB option, but you have many choices for both.

Here are some of the places you should consider for your computer-centric equipment (not all, but a good starting place):

- OfficeMax
- Office Depot
- Best Buy
- Circuit City
- Headsets.com
- HelloDirect.com
- AhernStore.com
- Communitech.com

USB handsets, being digital, are regularly touted as being higher quality and providing better sound quality. Honestly, I can't tell in the ones I've used, and many experts in the headset field call it a coin toss as well. The USB kind has only one plug to plug in, which may be an advantage to some extremely short-winded readers.

How do you choose? If your computer has a separate sound card, in place of or in addition to the speaker plugs on the motherboard, use an analog headset. The extra digital signal processing provided by the sound card will make a difference, although USB proponents say the outgoing sounds may be buffered too much and cause some jitter. A decent sound card in a recent computer should be plenty fast enough to eliminate this problem, whether it's a genuine concern or one manufactured by the USB crowd.

If you don't have a fairly recent computer or separate sound card, and have a choice of analog headset or USB headset in the price range you're comfortable with, try the USB headset first. Some experts disagree, but more lean toward USB.

Stop Mumbling

If you use speech recognition software, or are curious and want to try it, analog headsets will help. One of the strong recommendations by speech recognition software vendors is to use a quality microphone, and the headset in the medium and high range will do an excellent job.

Quick and Cheap (Less Than $30)

This category isn't the biggest, but you will have plenty of choices. Most lean toward the top of the range, but they're still less than 30 bucks (or so).

How inexpensive are some of these headsets today? You can find some, with brand names you will recognize, for $10 or so. That's pretty cheap, but you don't get much fidelity for that price.

What do you get for less than 30 bucks? Almost always, you get a single earphone for monaural sound and a fairly inexpensive microphone. You will probably have a volume control and maybe a mute button.

Figure 4-6 shows an affordable headset made by Plantronics, one of the pioneers in headsets of all types and all price ranges. Notice there's a single earphone. Notice it also includes a noise-canceling microphone, which may or may not work well–at this price point, you get what you pay for (not much).

Another key item, "Equipped with dual 3.5mm plugs for PC connectivity" is the next-to-last bullet point on the package. Look carefully when you examine headsets in person at the store or on a web site, because you must have two plugs and they must be 3.5mm in size. Headsets that look exactly like this may be designed to plug into a typical telephone

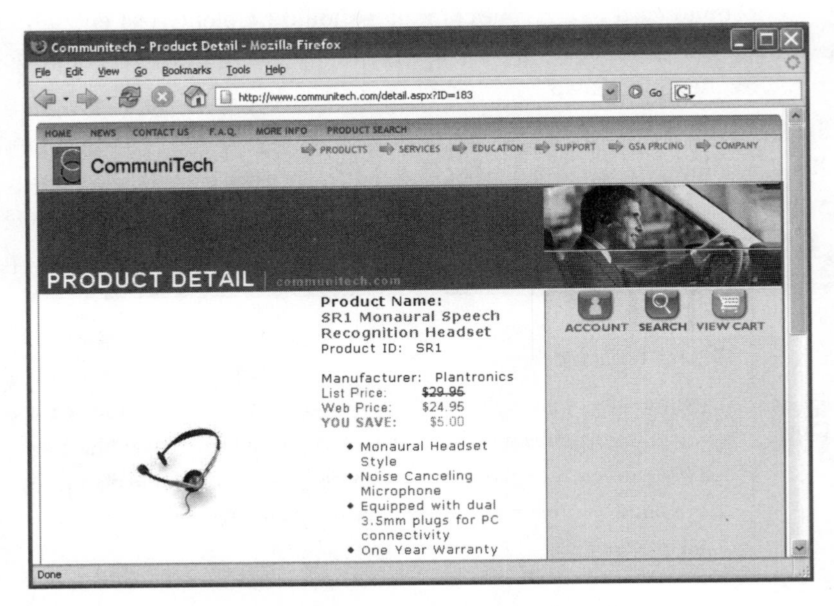

FIGURE 4-6. Under our budget and on sale besides

handset or base unit, making it easy to add a headphone to a standard phone. But those plugs are only 2.5mm, and there's only one plug. Be careful, because the two different types of headsets look amazingly similar when hanging in the store display.

If you want to go with the USB handset, you should. Many people feel more comfortable with a "real" phone to hang on to when using a computer-centric phone, and that's fine. The market has just recently grown to the point inexpensive USB handsets are available.

The least expensive handset I've found actually uses the two analog plugs for computer connection, even though all the other handsets I've found are USB. The $12.95 Riparius Internet Handset is for sale at *CommuniTech.com* (the company shown in Figure 4-6). In fact, they have a sale on now that reduces the price down to $7 for the handset. Go to Riparius.com for more details, but seven bucks is pretty inexpensive. Don't expect super sounds for seven bucks, but it is a physical phone that will make software phones more usable for many people.

Medium Quality, Medium Bucks ($30–$100)

This is the price range with the most choices and delivers a good balance of quality and price control. If you can't get a headset or computer handset you like in this area, you need your hearing aid checked.

Headsets in this price range start to feel confident enough about their quality to list frequency responses, just like stereo equipment. You will see many headsets advertise a 20 Hz to 20 kHz range. This will cover all the music you want as well as telephone calls.

If you travel and listen to music through your laptop, a headset with microphone for phone calling in this price ranges will deliver as good or better audio quality as dedicated music headphones costing about the same money. Why? Because the headphones are the basis for the headsets. Every headset lets you move the microphone on the boom up and out of the way, so you get both headset and music headphones (for audio or DVD movie soundtracks) all in one package.

Many offer two earphones, just like the headphones they are based on. This may feel weird for normal phone calling, but it helps if you're in a noisy environment. It also makes music sound like it does through normal headphones, because these are normal headphones. And if covering both ears bothers you, pull one earphone back off your ear when on the phone and it will feel more like what you are used to.

BestBuy.com offers a great comparison for three products in this price range. The page alignment is a little messed up, and their search process throws in a set of pure headphones between the two headsets, but that just illustrates my point about the music capabilities of these units.

Figure 4-7 shows that the two headsets offer two different styles of earcups. The Altec Lansing headset uses a closed earcup style, which blocks most outside sounds. These are great when you want to listen to music and ignore the world, and the process works the same for talking on the phone. If you're not familiar with the Altec Lansing name, they have a long and solid reputation in the music business as makers of speakers and headphones.

The Logitech headset doesn't offer the closed earcups, so you'll hear more outside sound while talking on the phone or listening to music. These may not seem as oppressive to some people who find closed earcups disconcerting. Logitech has a long and solid reputation making computer accessories, including speakers.

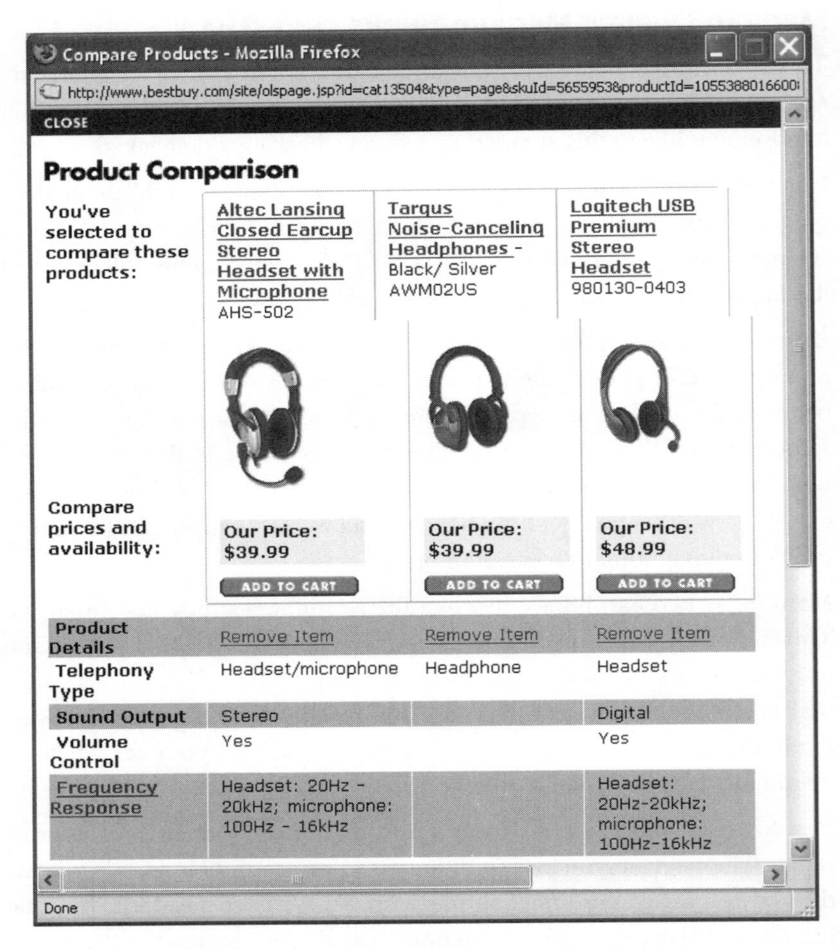

FIGURE 4-7. Two excellent choices for headsets, and a bonus

See the noise-canceling headphones in the center? They have closed earcups as well, making the case for choosing that style when sound isolation is important in your environment.

There are handset options in this price range as well. I have yet to find any USB handsets in the stores, as I mentioned, but you have multiple options online. For instance, go to PCMall.com and look for the Voice-Glo USB handset for $35.

Or you can get a combination handset/speakerphone with a nice futuristic style. Search for the Claritel i750H. You can start at Clarisys.com.

Notice in Figure 4-8 that you get a speakerphone as well with the Clarisys model. You can search for and find this product on two different online shopping sites (*www.communitech.com* and *www.CrystalVoiceLive.com*) in early 2005, and probably find it on more by the time your read this. As more people try software phones, more accessories like the Clarisys handset will appear on the market, and more retailers will sell them. You can already find handsets in the $10, $30, and $100 ranges, which indicates a growing market acceptance.

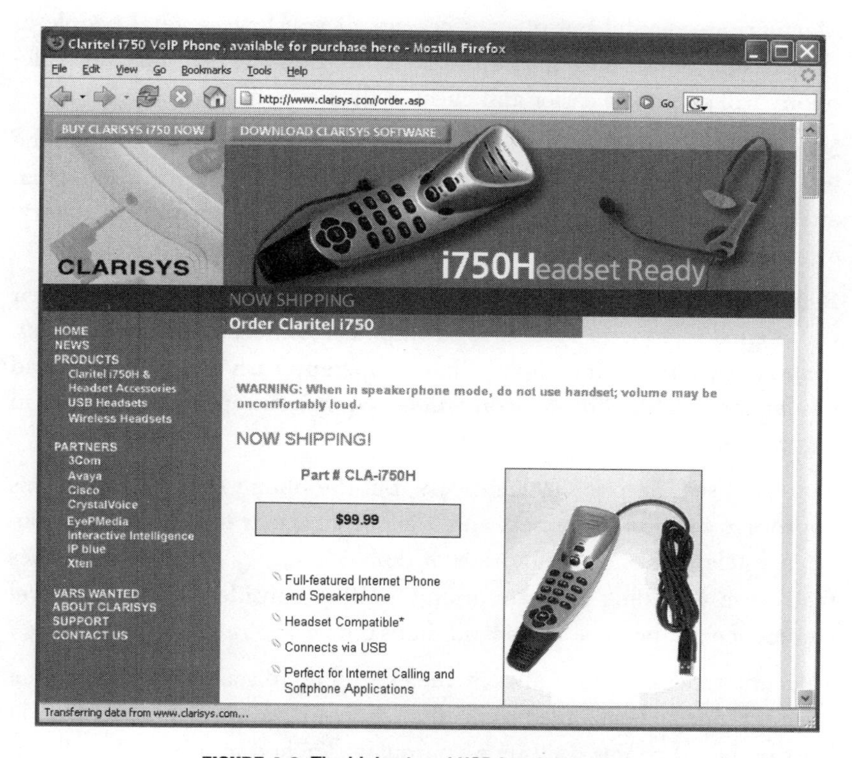

FIGURE 4-8. The highest-end USB handset available

High-End Voice Tools (More Than $100)

Can you spend more than $100 on a standard headset? Yes, especially when you look at products from specialized vendors like Plantronics (*www.plantronics.com*) who have a wide variety of wired and wireless headsets for every possible situation.

But can you take a leap up in productivity with broadband phones and make something that has been unaffordable suddenly affordable? Yes, when you are talking about audio conferencing equipment.

Teleconferences may be aggravating meetings, but they're better than driving across town or flying across the country for an aggravating face-to-face meeting. And with software phone services, especially Skype, you can have up to five people on a conference call and not pay a penny in teleconference charges.

Meetings are painful but often necessary. If you have several people in each of four offices and all need to be in on the same call, Internet Telephony will make this easier and cheaper than ever before.

Most conference telephones require an analog (traditional) telephone line. Most software phones have ways, through adapters, to connect an analog line to the computer running the softphone software. In theory, any teleconference phone can be connected to a softphone system.

But at least one company has jumped ahead and started advertising their conference telephone system for computer connection. Konftel (*www.konftel.com*), a Swedish company, has stepped up early in this market and has an affordable (for teleconference equipment) product, shown in Figure 4-9.

You can't see the price in Figure 4-9, but the phone sells for $349. This appears to be a lot of money, and it is–for a regular telephone. But conference telephones cost hundreds of dollars even for the miserable ones that cause unending frustration, and $349 is considered an entry-level conference phone. You can spend thousands if you're so inclined.

Teleconferences almost always mean a long distance call, and that means money. Teleconferences often happen in special rooms, or at least prepared rooms. Putting a computer attached to a broadband connection in a room and plugging in this conference telephone isn't much more trouble than just plugging in a telephone. There's even less trouble if you use a laptop with a wireless connection. That's as easy as putting down a box of donuts, and less fattening.

With Skype and your existing office-to-office data network, you have a teleconference system that costs no money for the software (Skype phone software is free) and no money for long distance charges (Skype-to-Skype calls over the Internet are free). You pay for the teleconference

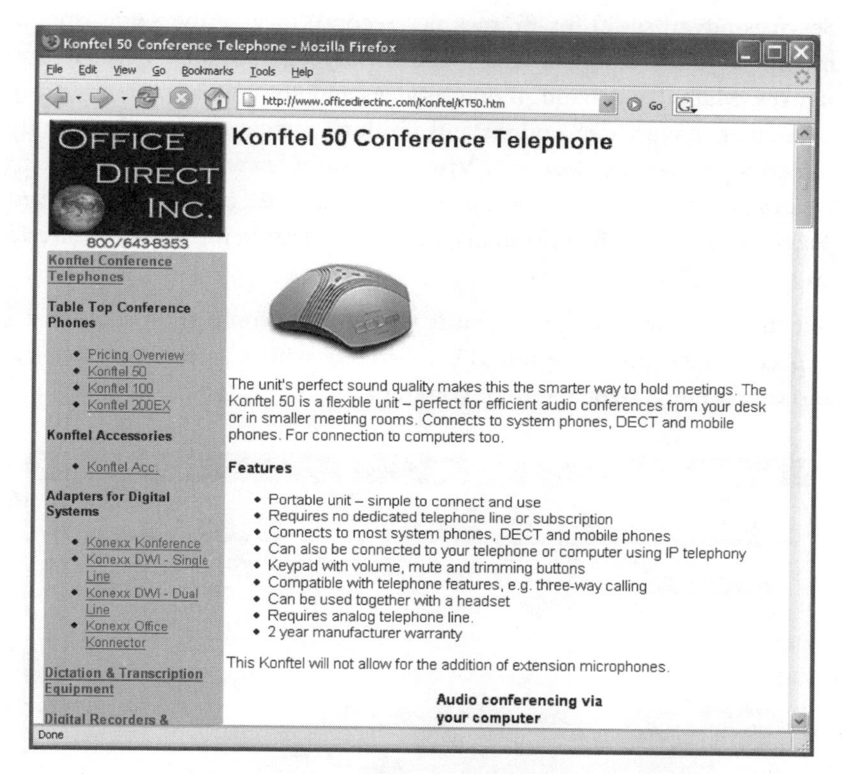

FIGURE 4-9. A US retailer for a new Swedish conference telephone

phone, but you'd have to do that anyway. And many times, you have to pay a third party, such as your major phone carrier, to set up a conference. Those calls are the furthest thing from free, and the long distance charges increment at every location, minute by expensive minute.

If you have a business with several locations and want to speak with many employees during meetings, a broadband phone system will make your life easier than ever. Check it out.

Beyond Standard Voice to Video

Don't get too excited, because the type of videophones you see on TV, with full motion support and incredible detail, aren't ready yet. However, if you miss someone and want to see their face, even if that face jerks around and the lips don't exactly sync with the sound, broadband video phones are ready.

Services advertise 30 fps (frames per second) to give the same smooth motion as television, but that's under only the absolute best circumstances. Many broadband phone services and computer accessory product makers have video options and products available. Vonage recently signed a partnership deal with Viseon (*www.viseon.com*). Logitech (*www.logitech.com*) has their VideoCall system ready to go. Polycom (*www.polycom.com*) has business systems that are affordable for small businesses and even individuals.

But the one I want to show you is the one that looks most like what I expect a video phone to look like: a phone with a screen and camera, shown in Figure 4-10. See what you think.

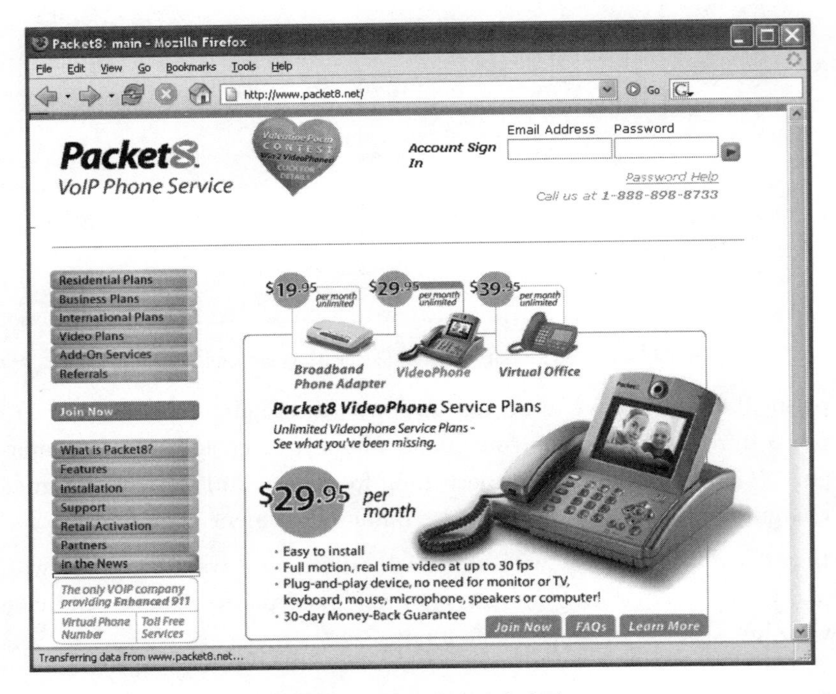

FIGURE 4-10. Cool phone, good price

You could argue a speaker phone would give a more natural look, and the unit does include an external microphone for just that purpose. You can also attach a larger display if you wish.

As I write this, the Packet8 videophone will only talk to other Packet8 videophones. Interoperability, or the ability for a video phone product from one vendor to work correctly with a similar product from a competing vendor, requires expensive equipment following high-end video-conferencing standards. These remain so expensive that many companies have avoided or abandoned their video-conferencing centers.

Don't let that stop you if you want to be a video phone pioneer. Check out Packet8 and the other companies flooding the market with web cameras and software to sync your headset voice to the image from the little camera sitting on your monitor. Look at this as an adventure.

Today, even bad video phone connections are impressive. So have fun and stick your tongue out at people telephonically.

Redial

You may not need any equipment when you switch to broadband phones, but if you do need some equipment, you have many choices in many price ranges. Don't let fear of new equipment stop you from switching.

5

VONAGE AND OTHER BROADBAND PHONE CARRIERS

Vonage leads the phone-centric pack of companies and has done the most marketing and advertising to create and expand broadband phone awareness among consumers. They deserve to be listed first, ahead of all their competitors, and so they are.

But remember that in many cases the features from one company closely match the features of their competitors. There are differences, and I will point some of them out when appropriate. But many times, when I say "Vonage," I could just as well say one of their competitors.

Watch this market attract more competitors as more consumers see the value of switching from traditional to broadband phones. Thousands of consumers per week switch in the United States, and that trend will continue. In fact, that trend will accelerate as competition drives prices down and features ever forward.

Companies aren't generic enough to lump them all into the "broadband phone company" bag and be done with it. Vonage does lead in consumer awareness and number of phone lines in use by those consumers, so they get more attention in this chapter. But Vonage can't look back, because something *is* gaining on them–their competitors. Those competitors range from new companies offering one service in the broadband phone market to all the Ma Bell descendants themselves moving into the broadband phone market to keep Vonage *et al* from stealing all their customers.

Hang on, because it's going to be a bumpy market for the next few years.

Broadband Phone Carriers

Here's how marketing works, at least on one level: if you've heard of the company, they will charge you more money than will a company you've never heard of. A name brand charges more than an unknown brand or store brand. You see this every day at the supermarket, where Del Monte charges more for a can of green beans than the store brand with more or less the same green beans in the same size can.

On the flip side of this, vendors know they must attract customers one way or another. Building a name brand and national reputation is one way to attract customers. Vonage accomplished this with their constant advertising on TV and all over the Internet. To many people, Vonage is the only broadband phone company, and Vonage loves that.

If name brands (like Del Monte and Vonage) mean quality and stability, the best choice of other vendors is lower prices. Every person has a threshold price where they believe the newcomer with a lower price is worth trying in place of the name brand with the higher price. You have to decide for yourself where that comfort point is for you.

One person helping you decide is Joseph Laszlo, an analyst studying broadband phones for JupiterResearch (*www.jupiterresearch.com*). "There are a significant number of players at the flat rate of $19.95 price point for coverage of all of North America. It's tough to make a national brand sell for less money." Based on this quote, it appears the price of a national brand name is about $5 per month per subscriber.

Consumers are telling me, by their purchasing habits, that an unknown brand must charge $20 or less rather than the $25 or $30 charged by well-known national players like Vonage and the major cable providers. Does that price difference make you feel comfortable? In essence, the unknown brands will let you keep the $5 per month they aren't spending on marketing to get your business. As a bonus, they get cash flow immediately when customers start to sign up, without having to wait until their company name reaches brand name status.

Chapter 2 lists a variety of Vonage competitors, but there are hundreds, believe it or not. For this chapter, mentally replace the word Vonage with "generic broadband phone service" and you'll be fine in many cases. When Vonage offers something much better, or much worse, than the majority of their competition, I will mention that fact.

Comparisons between providers can change when one of the companies updates their web site or drops their prices. Before you call or go online to order a broadband phone service, check again the features you need and want, as well as the current offerings of the service that's most appealing to you.

What You Get with a Broadband Phone

Every broadband phone company includes their list of standard features on their web site. They normally have a cleverly labeled menu item named Features. Figure 5-1 shows the Vonage Features page.

When you click on any of the features, a full page explanation of the feature loads automatically. The left side of the screen shows the list of features so you can easily drill down on any feature that strikes your fancy.

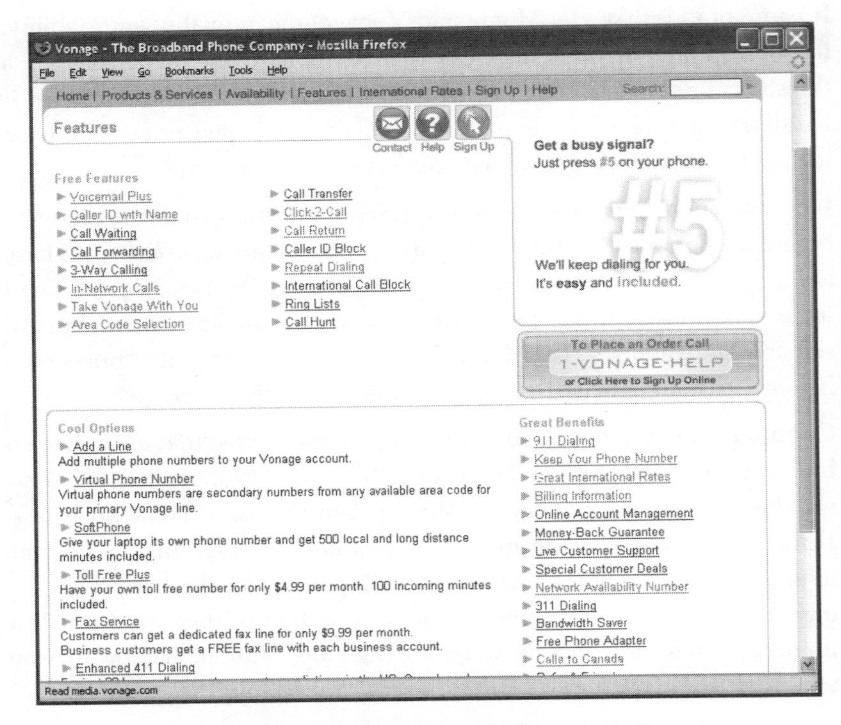

FIGURE 5-1. Main feature list on the Vonage web site

Some of the features you will always get from a broadband phone provider are:

- Free calls to other subscribers of the same service
- Caller ID
- Call Waiting
- Call Forwarding
- Make calls to traditional telephones
- Receive calls from traditional telephones
- Web-based control over your account and feature configuration
- Missed call notification/voicemail of some kind
- Calls to any phone in the continental U.S. without long distance charges

These are the minimum features you should expect from any phone-centric broadband phone service like Vonage. If you don't see these features on a potential service provider, click to the next broadband phone company.

Some differences between broadband phone providers can be determined by answering the following questions before signing up:

- Can you port your existing phone number to the new service?
- Is Canada considered part of free North American call zone?
- Are Alaska and Hawaii considered part of the free North American call zone?
- Is equipment provided to connect your phone to your broadband service?
- Can you to take your router on vacation and receive "home" calls there?
- What are the rates to foreign countries you call most often?

You have to evaluate this list carefully. For instance, there's a difference between a broadband phone company who doesn't offer a way to transfer your existing phone number to their service and an occasional problem with a local phone company who can't, or won't, transfer the number to broadband phone company. No matter how universal a feature is, there is some situation where a combination of technical problem and stubborn traditional phone company can cause you grief.

That said, check this list carefully. Some smaller broadband phone companies do not include the telephone adapter or router necessary to support your existing telephones. Some consider Alaska and Hawaii long distance calls.

Each provider prices international phone calls differently. Although these rates can change as partnerships between phone companies are made, pick a service provider who has a good rate to the countries you call most often.

This feature difference alone dictates services used by many new broadband phone customers. After all, if you have family in Ireland, you want a service with cheap calls to Ireland. The fact that they charge three times as much as a competitor for calls to Australia won't matter if you don't call Australia.

Standard Features

Here's the list of standard (no additional charge) features from Vonage as of early 2005. Many of their competitors offer the same or similar features:

Voicemail Plus
: Get voicemail messages by phone or web page, or have them emailed to you.

Caller ID with Name
: Caller's name appears on your phone and in the subject line of any voicemail messages emailed to you.

Call Waiting
: Juggle two calls at once (don't get confused).

Call Forwarding
: Forward calls from one number to any other number, or have both phones ring at the same time (their free Simul-Ring feature).

3-Way Calling
: Call and talk to two people together in one conversation.

In-Network Calls
: Calls to another Vonage customer don't count in your monthly minutes allowance.

Take Vonage with You
: Take your telephone adapter (or router) and a telephone, and you can receive your home calls anywhere you have Internet access. You can take your router (and number) with you when you move as well.

Area Code Selection
: Your Vonage phone can have any area code from any Vonage service area you want.

Call Transfer
: Transfer any call to any other phone number.

Click-2-Call
: Click a phone number on your computer, such as one in an address book, and your phone will ring and immediately ring the number you clicked.

Call Return
: Use the *69 service to return calls you missed.

Caller ID Block

Use the *67 service to stop your name from appearing on the receiving end's Caller ID screen.

Repeat Dialing

Push two buttons to have a busy number called constantly for up to 30 minutes.

International Call Block

Stop anyone from dialing international calls from a number, and directory service too.

Ring Lists

Choose different ring tones when you have more than two or more Vonage accounts and numbers connected to a single telephone.

Call Hunt

Automatically ring other Vonage numbers on your account when the primary number is busy. You can route the hunt sequence over 10 numbers.

Will you use all these features? No. Will some potential customers feel one of these features is critical or they won't sign with the service? Absolutely.

One way to differentiate your service, even without becoming a brand name, is to offer more features than the market leader. In this market, that means offering more than Vonage.

Both BroadVoice and Packet8, for example, offer better Caller ID blocking than Vonage advertises. If you want Caller ID blocking on all calls rather than setting that option call by call with *67, check out BroadVoice and Packet8 (among others). That's a minor difference, but if you have strong preferences for features, then check the feature list carefully.

Besides features, money helps many people decide on a provider. Do you want the cheapest possible monthly payment? Most do. Do you feel comfortable paying up to $30 for an installation fee to save $5 to $8 per month on your bill? Do you want to own your router or telephone adapter, or get one from your service? If you buy your own from a third party, you get a monthly discount and you have a better chance of using that equipment with another vendor if you switch services.

Surprise Restriction

Routers and telephone adapters provided by broadband phone companies are almost always keyed to that provider only, and WILL NOT work on other services.

On one hand, everyone wants the best deal possible. On the other, picking a service based strictly on a couple dollars per month is pretty shortsighted. If you feel more at home with the web site, information pages, feature lists, and support pages of one vendor, that's worth a dollar or three per month. Saving a dollar while losing your patience with a provider that's a poor fit is not worth the aggravation.

Advanced Features

Broadband phone companies call the features listed earlier their "standard" features; and they also offer "advanced" features. Personally, I believe if a feature comes as part of the package price, it's standard. However, all the features listed as standard are comparable to those offered by traditional telephone companies.

Marketing advice probably convinced the broadband phone companies to call some of their features "advanced" to mark the differences between the new wave of telephony (broadband) versus the old wave (Ma Bell). Take a look at these features from BroadVoice and see if any of them have been offered by your traditional telephone company:

Call Manager
: Web-based account configuration screens.

Calling Line ID Blocking per Call
: You can block Caller ID information on all calls through the configuration screen, or call by call with this feature.

Call Forwarding Selective
: Normal call forwarding sends all received calls to a different number, but this lets you forward calls from certain numbers to specified numbers. BroadVoice allows you to specify individual or ranges of phone numbers and forward those numbers differently.

Calling Line ID Delivery per Call

If you block Caller ID information generally, you can enter *65 before dialing to send your Caller ID information on that call only.

Priority Alert

Define a special ring for special callers or phone numbers to alert you aurally when they call.

Now you may think, as I do, that the Call Manager or Online Account Management screens are a standard part of broadband phones, since you have to use these to set up and manage your account. But your traditional telephone company doesn't have anything like this, do they? That's why the broadband phone companies make a big deal about it.

BroadVoice puts their Call Manager under the Advanced feature set. Vonage lists their Online Account Management under Great Benefits. Packet8 also prefers that you pay your bill online, to save time and stamps, through their Online Management screen.

Here are some features that just Vonage offers (in most cases), listed under their Great Benefits heading:

311 Dialing

Reach city nonemergency numbers through Vonage as you can with your traditional telephone today. Most broadband phone providers don't offer this.

Network Availability Number

If your Internet connection goes down completely (it happens much less often than in the past, but it does happen), Vonage will transfer all your calls to another specified number, such as your cell phone, traditional phone line, or another person. (VoicePulse now offers this service, which they call Line Unavailable Forwarding.)

911 Dialing

Not exactly the same as the 911 that the telephone company provides, but close. (See "911 Support," later in this chapter).

Refer-A-Friend

Convince a friend to sign up for your broadband phone service, and your service will reward you. Check out Table 5-1 for more details.

VoicePulse offers Telemarketer Block, which may be worth changing your phone service for, all by itself. Lingo offers Automatic Call Rejection, which refuses calls with numbers blocked out or listed as anonymous, common tricks of telemarketers. Packet8 offers call blocking of anonymous calls as well.

One way for companies to get more business is to encourage their happy customers to become salespeople. Car dealers call these "bird dog fees" (at least in Texas) when you send them a new customer. For the hunting-impaired, bird dogs flush out the game, and you're flushing out new customers for the salesperson.

The fees can be worthwhile, as you can see in Table 5-1.

TABLE 5-1. Customer referral fees

Company	You get:	Your referrals get:
Vonage	Two free months (up to $50)	One free month of any service plan, business plans included
Lingo	$25 credit	$25 credit
Packet8	One free month	One free month

I'm frankly surprised more services don't offer rewards for signing up customers. Every customer you give them is another customer they don't have to find on their own, and customer acquisition is one of their major costs. Maybe some other services will get smart. In the meantime, if you have a wide group of friends who are easily influenced and interested in switching their telephone service, you can call free forever. Family members qualify, too, so harangue those cousins.

Optional ($$) Features

Not every possible telephone service comes included in your monthly broadband phone subscription. There are several features that cost more money, but they cost less than comparable services from your traditional telephone company. Assuming your Ma Bell leftover has these services at all, of course.

Add a Line

Run two or more phone lines through your Vonage router(s).

Virtual Phone Number

Add a phone number from a different area code to allow callers in that area to make a local call to you.

SoftPhone

Add a software phone to your PC, Macintosh, or laptop.

Toll Free Plus

Have your own toll-free 800 number.

Fax Service

Use the second phone plug on your router for a fax machine. Business customers get the fax service as part of their standard package.

Enhanced 411 Dialing

Get directory information from all over the U.S., Canada, and Puerto Rico.

Can your current telephone company give you a telephone number in a remote area code? Actually, they offer that service for businesses, but it costs much, much more than the $5 per month Vonage charges.

Other broadband phone providers have almost the exact same feature list, but they also have a couple of extras that Vonage doesn't yet include. For example:

Video phone service

Packet8.

Music on hold

BroadVoice.

Virtual phone numbers in other countries

Lingo offers virtual numbers in over a dozen countries; Vonage offers them in the U.K., Canada, and Mexico; and BroadVoice offers numbers in the U.K.

Microsoft Windows Instant Messenger link

BroadVoice.

In addition, every residential broadband phone provider offers business services. If you have a small company, broadband phone services will save you a great deal of money and provide more features than you could ever afford from a traditional telephone company.

Pricing for these optional features varies greatly between broadband phone providers. Some nice charts for standard services and optional features are waiting in the "What It Costs" section later in this chapter.

911 Support

On one hand, people have come to rely on 911 and feel it's somehow a basic right. On the other hand, people did perfectly well for the hundred years or so before 911 became available when they had to call the police directly.

Please realize that the Ma Bell leftovers constantly yell about 911 service, even to the point of pushing stories to local TV news shows when a tragedy occurs in a home that has broadband phones but no 911 service. These stories are an example of A) corporations concerned about the safety of their customers or B) greedy corporations fighting competitors using fear rather than innovation. This isn't a test: you decide which you feel is correct.

However, let me say that this worry about 911 calling will disappear in short order. Here's what the broadband phone companies will be able to do when city emergency systems catch up:

- Track the serial number (MAC address) of the router that you use with your broadband phone service.
- Put that serial number in a database with your name and address.
- Grab 911 calls and route them to the local emergency contact point for your location.
- Send your name and address from their customer database directly to the 911 dispatcher computer system.

These features will match what you have with traditional telephone line 911 today. In another year or two, however, people will start to point the finger of inadequacy toward the traditional telephone line 911 services because of the advances of broadband phone services. Here are a few emergency features broadband phones could provide in the next year:

- Send floor plans of the location to the 911 operator (gathered from apartment buildings or home builders providing schematics).
- Tie in medical information by pulling a link to your medical records on file at your doctor and send that info along with the emergency call.
- Send photos of family members living at your location to emergency services.
- Immediately fax any medical authorization forms to the nearest hospital if necessary.

These are obvious ideas derived from common sense and a few minutes' thought. Chapter 7 explores some of the advanced telephone services that are becoming available thanks to broadband phone technology.

Chapter 7 also has the full scoop on 911 services available through broadband phone companies. Rest assured Vonage is at the leading edge of 911 capabilities available today from broadband phone service providers.

Comparing Providers

I can't look at all the possible broadband service providers in this section, because there are hundreds, and this is a small book. I can show you what Vonage offers and how several of the major competitors to Vonage stack up. Use this list as a guide to see which features are important and which ones aren't.

I picked features that many users report influenced their decisions when choosing a broadband phone service. The features in Table 5-2 are not listed in order of importance, because only you can prioritize what you want from your new phone service.

TABLE 5-2. Quick feature comparison (subject to change)

Feature	Vonage	Packet8	Lingo	BroadVoice	ATT CallVantage	VoicePulse
911 support[a]	Yes	Optional	Yes	No	Yes	No
Keep your number?	Yes	Yes	Yes	Yes	Yes	Yes
Activation fee	Yes	Yes	Yes	Yes	Yes	Depends on plan
Speed dial	No	No	Yes	Yes	Yes	Yes
Band-width saver	Yes	No	No	No	No	Yes
Soft-phone support	Yes	No	No	No	No	Yes

[a] As of 6/1/2005. By the end of 2005, all broadband phone services must provide 911 service.

Some features (or options) are consistent across all broadband phone providers. Others vary, but you may not care. Vonage may not have speed dial, but does that matter if you already have your speed dial numbers programmed on your phone? Probably not.

Bandwidth saver won't come into play all that often. But if you have two or more broadband phones that can be used at once on a slow broadband connection, limiting the bandwidth so that all concurrent calls will be usable will be important to you.

The crucial part is deciding on the features you need and checking with the service providers to make sure they support those features. Feel free to pick up the phone and call their toll-free numbers to see whether they answer your question quickly and competently.

Before You Sign Up

If you're sold on the idea of going with a broadband phone provider, you're almost ready to take the plunge. The next few sections will get you the rest of the way there.

What You Need

What are the equipment and service requirements you need to provide in order to sign up for a broadband phone service? Not many, because the service provider will send you the correct router or telephone adapter for your situation.

Your requirements are:

- Broadband service with at least 90 kbps upstream speed per active phone to be used concurrently. (Your broadband provider can tell you your upstream speed, but the slowest broadband service provides 128 kbps upstream and does fine with a single conversation.)
- A computer for system configuration and modifications.
- A valid credit card to cover the monthly service fee.
- A shipping address (not a P.O. Box) to receive the equipment from the service provider.
- An analog phone to connect to the router or telephone adapter.

Notice that even these requirements aren't that onerous or binding. The slowest broadband service available today, entry-level DSL, provides at least 128 kbps upstream bandwidth. That will support one broadband phone call with excellent voice quality, although heavy Internet activity at the same time may cause some degradation.

Could you sign up a child who's living at college and have your credit card billed? Certainly. Can you use any old telephone to make broadband phone calls? Absolutely.

There's one more requirement: the desire to save money or increase the number of features your phone has by switching from a traditional telephone line to a broadband phone. But that's an easy requirement to meet.

What It Costs

How much a service costs is always an important question. The fast answer about broadband phones? Less than you're paying for a traditional telephone line. Sometimes *way* less than you're currently paying per month for your traditional telephone.

Prices will change (drop) over time as the competition heats up and more people start switching to broadband phones. For this reason, I'm hesitant to list prices here, a book that can't change after printing to reflect dropping prices. However, the value of comparison shopping outweighs my nervousness about pricing changes. But let me list some caveats:

- These prices will change.
- Higher-priced vendors will drop their prices, but probably stay higher priced than the competition.
- Cable companies and current traditional telephone companies don't yet understand how to price these services competitively.
- Lower-priced vendors will probably stay under the price of the name brand vendors.
- Consolidation will absorb one or more of these companies, possibly before this book even gets printed.

Hesitation aside, take a look at Table 5-3 for an indication of which service sets their prices at which levels.

TABLE 5-3. Broadband phone service pricing (subject to change)

Vendor	Basic package	Premium package	Activation fee	Shipping
Vonage	$14.99	$24.99	$29.99	$9.96
Packet8	$19.95	$19.95	$29.95	$0.00
Lingo	$14.95	$19.95	$29.95	$9.95

TABLE 5-3. Broadband phone service pricing (subject to change) (continued)

Vendor	Basic package	Premium package	Activation fee	Shipping
BroadVoice	$9.95 (no long distance included)	$19.95	$39.95	$14.95
AT&T Call-Vantage	$19.99	$29.99	$29.99	varies
VoicePulse	$14.99	$25.99	$29.99	$9.99
TimeWarner	$39.95	$39.95	$0.00	$0.00

Notice who rates their service worth more than the other providers? The old telephone company (AT&T) and one of the major cable companies (Time Warner). I wonder how they get customers. Just another reason you should always comparison shop.

The majority of the broadband phone providers offer comparable rates (so close they may be setting their prices based on competitor rates rather than their real costs, but that's a subject for a business book). If the services you like charge about the same, decide whether you want any of the optional features. If so, take a look at Table 5-4 to make sure your favorite provider offers the services you want at a reasonable price.

TABLE 5-4. Optional services pricing (subject to change)

Vendor	Alternate/virtual numbers	Toll-free virtual numbers	Fax line	Enhanced directory services
Vonage	$4.99 per month	$4.99 per month, 4.9 cents per minute after 100 minutes	$9.99 per month, $9.99 activation	$.99 per call
Packet8	$4.95 per month, $9.95 activation	$4.95 per month, $9.95 activation, 3.9 cents per minute after 100 minutes	n/a	$.75 per call
Lingo	$4.95	$10.00	n/a	$.75
BroadVoice	$1.95 per month, $9.95 activation	$1.95 per month, $9.95 activation, 2 cents per minute after 60 minutes	n/a	n/a

TABLE 5-4. Optional services pricing (subject to change) (continued)

Vendor	Alternate/virtual numbers	Toll-free virtual numbers	Fax line	Enhanced directory services
ATT Call-Vantage	$4.99 per month	n/a	n/a	n/a
VoicePulse	$5.14 per month, $7.99 activation	n/a	n/a	n/a

You can see that providers do offer some choices with their special packages, although it's hard to say with a straight face that some services provide anything "special." Again, the new entrants into the market offer the best services and best pricing, and Vonage leads the way in both (or it's darn close).

Some broadband phone providers are starting to offer unlimited calling plans to certain parts of the world. Check your provider to see whether they add an unlimited calling plan to an area you wish to include in your calling plan. For example:

- Packet8 adds Europe or Asia to their unlimited North American calling plans for a monthly price of $49.90. You can get unlimited calls to both areas for $79.90 per month.
- Lingo offers unlimited calls to Asia for $39.95 per month and unlimited international calls for $79.95 per month.
- BroadVoice offers unlimited calling to 35 countries for $24.95 per month.

More unlimited calling to countries in Europe and Asia will be the trend as the supporting connections between the broadband phone services and local telephone networks in those countries increase with time. Sad to say, but the idea of a special "long distance" telephone company doing nothing but long distance call routing died when SBC bought AT&T in February 2005. Before long, a call will just be a call, no matter which countries are involved. This is good for consumers, of course, unless they had a lot of retirement money in long distance phone company stocks.

Currently, there is only a small amount of federal tax on broadband phone service. Most plans include less than $2 of taxes and fees assessed by various governments, although some states charge sales tax on these services. While your traditional telephone line providers double the cost of your monthly payments through assorted taxes, fees, and surcharges, you won't see that on your broadband phone bill.

Decision Checklist for New Users

Switching to a new phone service provider can be a bit nerve-racking. Remembering all the features you want to get on your new phone service can be tough when you keep getting distracted by this price and that price and by feature pages from competitors that look far too similar to be an accident.

Let me help you define what's important for your new broadband phone service. Go through the following table and choose Very Important, Somewhat Important, or Not Important for each item. When you finish, keep that list beside you when comparing broadband phone providers.

The order doesn't really matter. The weight you give to each feature will be different than what your neighbor gives, and that's okay. Your goal is a phone service that fits what you need.

This is not a test. There are no grades. No one else will see your answers. The more honest you are with yourself about what you really want from your phone service, the better service selection you will make.

You might want to use a pencil so you can change your mind and your answers. For all of the following statements, assign criteria of Very Important, Somewhat Important, or Not Important:

Criteria	Very important	Somewhat important	Not important
I need to keep my traditional telephone line.	❏	❏	❏
I need to keep my same phone number.	❏	❏	❏
I need my phone number to be listed in the white pages.	❏	❏	❏
I make a lot of long distance calls within the U.S.	❏	❏	❏
I make a lot of calls to Europe.	❏	❏	❏
I make a lot of calls to Asia.	❏	❏	❏
I want to save money.	❏	❏	❏
I need modern features like voicemail delivered via email.	❏	❏	❏
I need 911 service.	❏	❏	❏

Criteria	Very important	Somewhat important	Not important
I use directory service operators regularly.	❑	❑	❑
When I need support, I want to call and talk to a person.	❑	❑	❑
I need the lowest monthly charge possible.	❑	❑	❑
I need the lowest upfront cost possible.	❑	❑	❑
I want to choose my router or telephone adapter at a retail store.	❑	❑	❑
I need a softphone as part of my service for when I travel with my laptop.	❑p	❑	❑
I need a provider that's well established.	❑	❑	❑
People from all over the country need to call me for free.	❑	❑	❑
I have friends I can convince to join the service, so I need a referral program.	❑	❑	❑
It needs to be easy for friends and family from back home to call me.	❑	❑	❑

How can these answers help you choose your broadband phone service provider? There are two steps needed before we can answer that question together.

First, prioritize the items you marked as Very Important. You may have marked contradictory items, such as asking for the lowest price but also for a well-established provider with live telephone support. (I want a car with 400 horsepower and room for 5 adults that gets 40 miles per gallon and costs under $15,000. I am perpetually disappointed, but I don't want you to be disappointed when you switch your telephone service.) Go through and put numbers from 1 through 5 by the items you feel most strongly about and checked Very Important, ranking them from most important to least important (least important of the important items, but you know what I mean).

Second, decide whether any one item is absolutely a deal breaker. If you must keep your current phone number, for instance, that requirement must be Number 1 and will be at the top of your list. Or, if you need to keep your listing in the phone book, you should know that broadband phones don't get listed–although if you keep your current phone number, it will stay listed for a year or two.

It doesn't matter what your requirement is, but if there is one item that you must have, be honest and mark that item. If there's no deal breaker, then don't worry about it.

Now take your Top 5 list and compare the items to provider web sites. Start with Vonage, because they are the focus of this chapter, but continue on through the other providers that meet your requirements. If you're having trouble getting the match you want, don't limit yourself to the few listed here in the book. Go to any search engine, type in "broadband phone service provider," and you will see scores to hundreds of results. Or check out a portal like *www.broadbandreports.com/isplist?t=voip*, or my web site at *www.gaskin.com/talk* for this information.

Signing Up

Does it seem obvious that a phone company would provide a toll-free number on their home page so you could call and sign up for their service? It does to me, but only Vonage and Packet8 make their toll-free numbers obvious.

Of course, you have to have broadband and an Internet connection before any of the broadband phone services can help you, so expecting you to sign up online isn't that great a stretch. Let me walk you through what will happen when you go online to sign up, using Vonage as an example:

1. Choose Sign Up from the top menu or Retail Activation from the bottom right of the home page.
2. Choose your service plan (Residential Premium or Basic, Business Premium or Basic).
3. Choose to get a new phone number from Vonage (or transfer your old number on this screen).
4. Choose the state, area code, and region for your Vonage phone number.
5. Provide personal information for billing and router shipment (you can't use a P.O. Box).
6. Agree to the Vonage terms and conditions.
7. Put in your address information and choose a Vonage username and password. Your password must be eight characters long with no spaces. Pick a mix of letters and numbers without any real words inside for better security ("password3" is not a good idea).

8. Provide credit card and billing information.
9. Verify the order, and then hit Order Now. Your initial order charge will be shown on this screen.
10. Wait for your new phone equipment to arrive.

There is less trouble ordering a new phone service from Vonage than on many e-commerce web sites. Everything is explained clearly, as you'll see in the following screen shots, and web pages can change at any time. Let me show you a few of the pages so you'll get a flavor of how they look and what type of information you must provide. Figure 5-2 shows the first page you see after clicking on the Sign Up menu item.

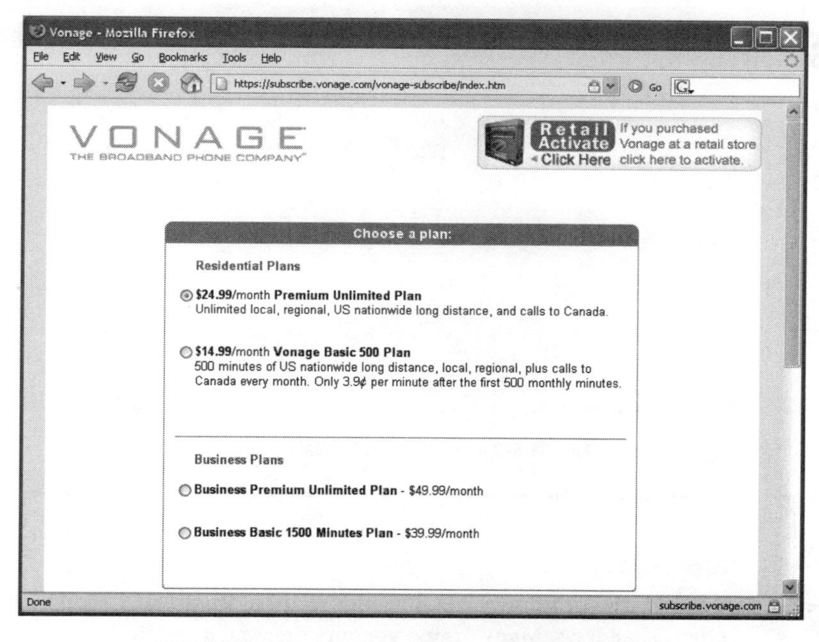

FIGURE 5-2. Half of the first page you see when signing up

Notice the little lock sign in the bottom right corner of the screen? That shows the Vonage web page is using a secure connection. All the information you provide, especially financial details like your credit card number, will be encrypted before going across the Internet.

Notice the blue box in the top right of the page? If you bought a router at a retail outlet or online, click that link to start the activation process. You can also reach those screens from the home page as well.

Figure 5-3 shows the interesting option you have with virtual numbers. Although I'm in Texas, I'm selecting a phone number that's local to Cleveland, Ohio. Why? Because I can, and because someone in Cleveland may want to call me, but not enough to pay the long distance charges (obviously they're not on Vonage yet).

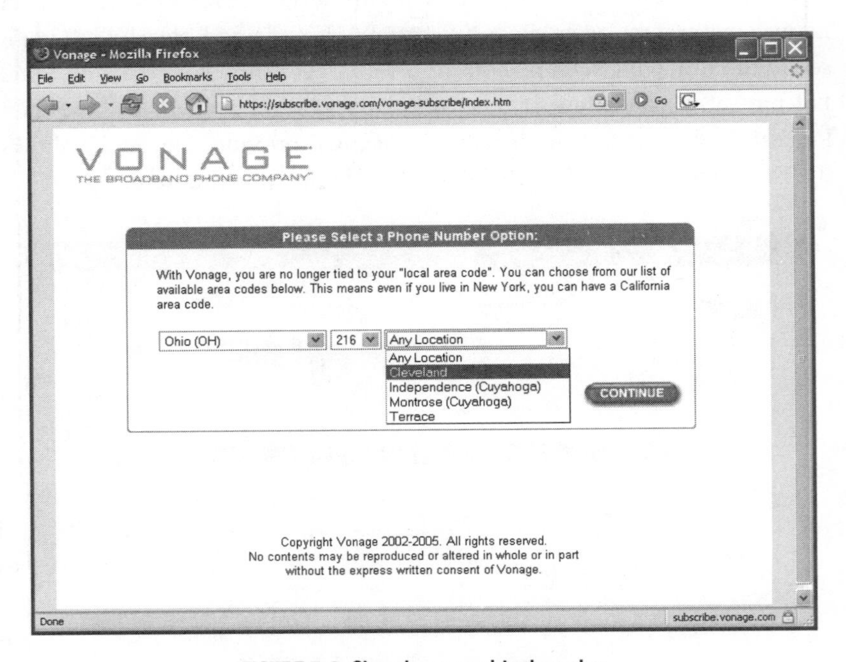

FIGURE 5-3. Choosing your virtual number

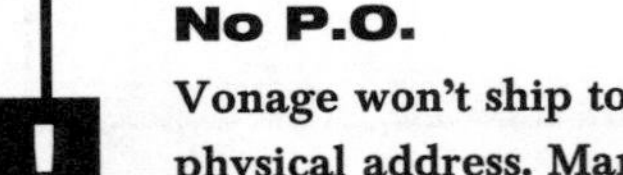

No P.O.

Vonage won't ship to a P.O. box so you must provide a physical address. Many online vendors demand a physical address to cut down on fraud.

When you get to the last screen, you will see the total for all your activation, shipping, Federal Excise Tax (FET), and the Regulatory Recovery Fee (the government always gets theirs somehow, don't they?). The total shown in Figure 5-4 is what will appear on your first month's billing.

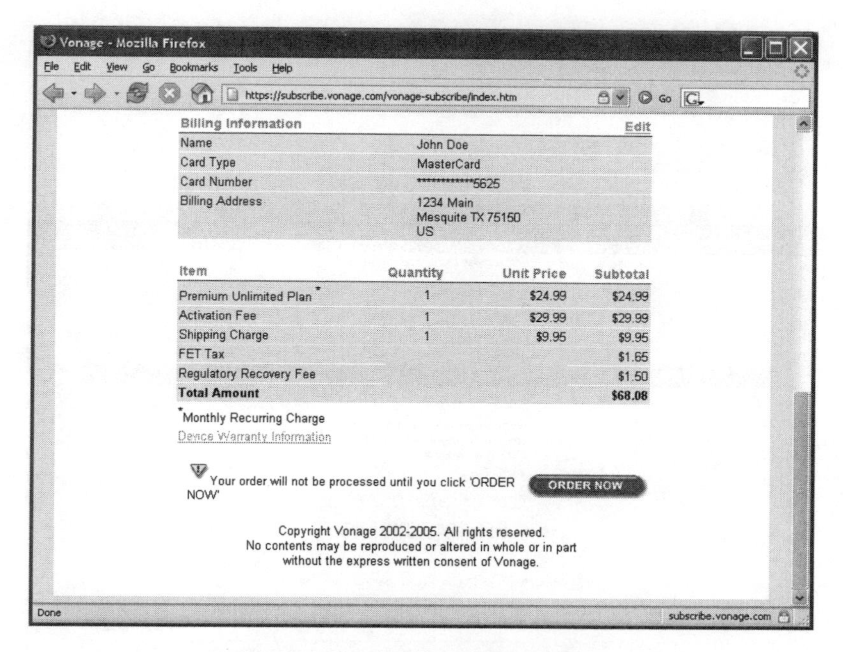

FIGURE 5-4. Your verification page before ordering

Obviously, your monthly billing amount will not include shipping and activation charges. Every broadband phone service company charges similar fees, except for occasional specials for new users.

If you have a friend with Vonage already, let them refer you. That will cut one month's service fee off one of your early bills, and your friend gets up to a $50 credit. Then make them take you to lunch.

When you purchase a router before signing up for Vonage, the screens differ somewhat. One major difference will be registering the router or telephone adapter you purchase that comes with the Vonage service. There will be instructions in the box, but before you start your sign-up process, make sure you have the equipment box with you. Figure 5-5 shows the screen asking for the MAC (Media Access Control) address of the hardware you purchased.

Before you see this screen, you will tell Vonage whether this is a new activation of the Vonage service or if you're adding a line to existing service. After this screen, things progress pretty much like the earlier steps and screens.

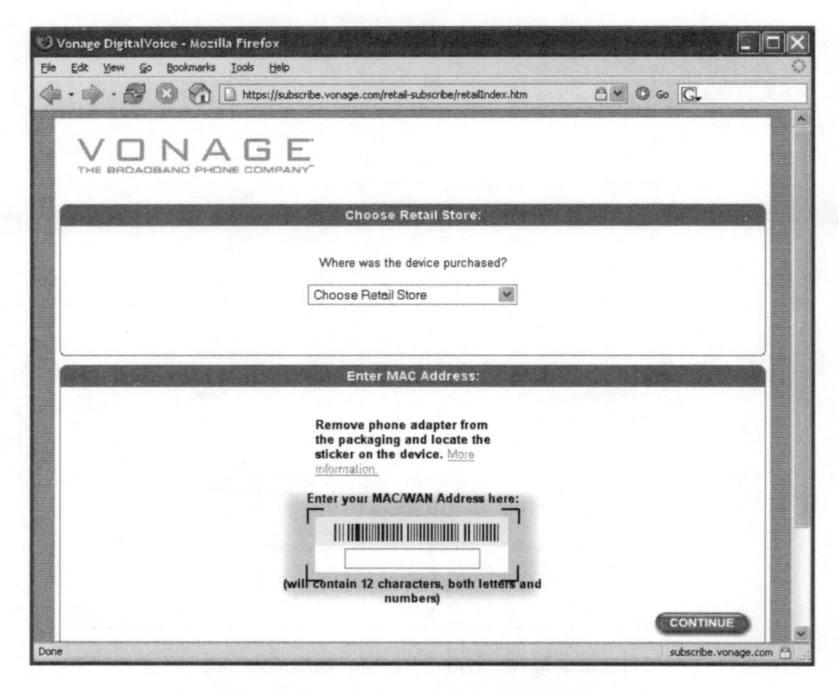

FIGURE 5-5. Signing up with pre-purchased hardware

One interesting note is the pull-down menu that lists where you purchased your Vonage device. Early in 2005, the screen already listed 67 different retail and online chains. Any question why Vonage is the market leader when they sign up that many retail partners?

> **!**
>
> **DSL Users Beware**
>
> **Do not transfer your telephone number to Vonage if your broadband connection is a DSL connection on that number. Broadband providers are working on ways to provide what they call *naked DSL*, which is a phone line with DSL but no phone. This is available in only few places today.**

You can actually come out ahead financially by dropping your traditional telephone service options down to the bare minimum and adding a broadband phone to that line. The combination cost of the broadband phone, especially if you get a metered service for $10 or $15 that provides all the regular features but only a limited amount of calling minutes, will be less than a traditional phone line with Caller ID and any other service.

Signing up with other broadband phone services works much like the Vonage example here, because they will need similar information. You'll have to provide an address, choose your virtual number if you want one, and provide your credit card information for billing.

After You Sign Up

The first thing you'll probably do after you sign up is make a few phone calls. Once you have that out of the way, there are some things you need to know how to do.

Monitoring Your Account

Every broadband phone company uses a web configuration and administration utility so customers can monitor and update their service at anytime from anywhere. If you remember waiting for your traditional telephone company to respond when you wanted to make account changes, you'll be thrilled with how easy and quick such changes happen with broadband phone service providers.

Using a web administration interface means that yes, you can control the account details for a family member if they give you their username and password. You can change your account settings or check on your account from anywhere you have access to a web browser, such as a coffee shop around town or an airport halfway around the world.

The Vonage folks do an excellent job with their Dashboard view of all critical account settings and call details on one screen. Other companies have similar offerings, but I will use Vonage to illustrate the types of administration screens you will see on all the services.

Vonage's Dashboard

The Dashboard appears first when you log into your Vonage account. As you can see from Figure 5-6, just about everything you want to do with your account can be launched from this screen.

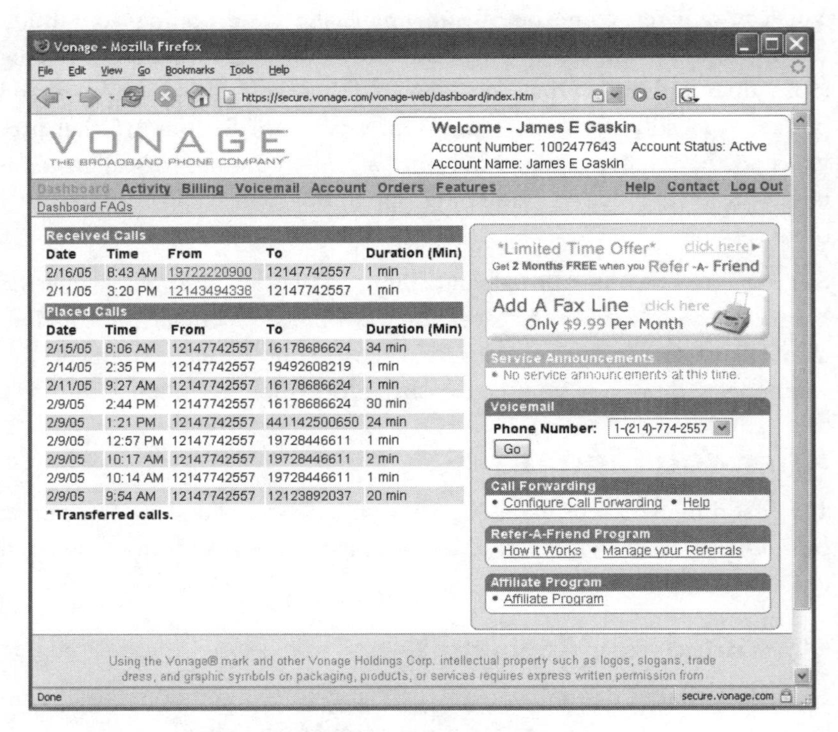

FIGURE 5-6. The Vonage dashboard

You can configure your startup screen to go to a different view from the login screen. For example, if you retrieve your voicemail over the Web often, going straight to the voicemail screen would save you one click. Notice the Voicemail section in the middle right of the screen, where you can press the Go button and check voicemail for the number listed. If you have more than one Vonage phone number, use the pull-down menu to define which voicemail box to check.

Also notice that the shortest call duration is "1 min" for one minute. Since you aren't charged for incoming calls with Vonage, there's no reason for them to slice a minute into smaller pieces for accurate billing like a traditional long distance company does to more accurately charge you for the exact time used. The top call listed actually left a 13-second voicemail message (you'll see that in the next section), but appears to be a 1-minute call.

When you put your cursor over the Received Calls From listings, Caller ID information appears. Not much information, but enough to remind you who called if you don't recognize the number.

The second option listed in the horizontal menu under the Vonage name is Activity. This screen just duplicates the call information with little added functionality, so I won't waste your time showing it to you.

Catch the Sales Pitch

Vonage never misses a marketing opportunity, and you can see four sales pitches on the screen in Figure 5-6. First, the "Limited Time Offer" to get two months of free service when you refer a friend. A second Refer-A-Friend plug is lower on the right side of the screen.

The third sales pitch is to add a fax line for only $9.99 per month. That's a discount from the basic $14.99 or premium $24.99 monthly fee, so heavy fax users could save some money with this option by avoiding the long distance charges that they would incur with a traditional phone line. Finally, the Affiliate Program is another type of customer referral award program.

But this is okay with me, because I believe only a few of the hundreds of broadband phone providers will maintain a national presence. Vonage will be one of them because of their marketing savvy, although some may call it marketing saturation. Besides, a personal referral from someone you trust is about the strongest sales pitch you can get, and Vonage encourages you to share your broadband phone happiness. Constantly.

Voicemail management

The second most visited web administration page, at least for me, is the voicemail page. Figure 5-7 shows my page with one short message.

Notice you have a pull-down menu to set the current phone number for the voicemail check. If you have multiple numbers, you have to remember to check them all, unless you received an email already about the message in question.

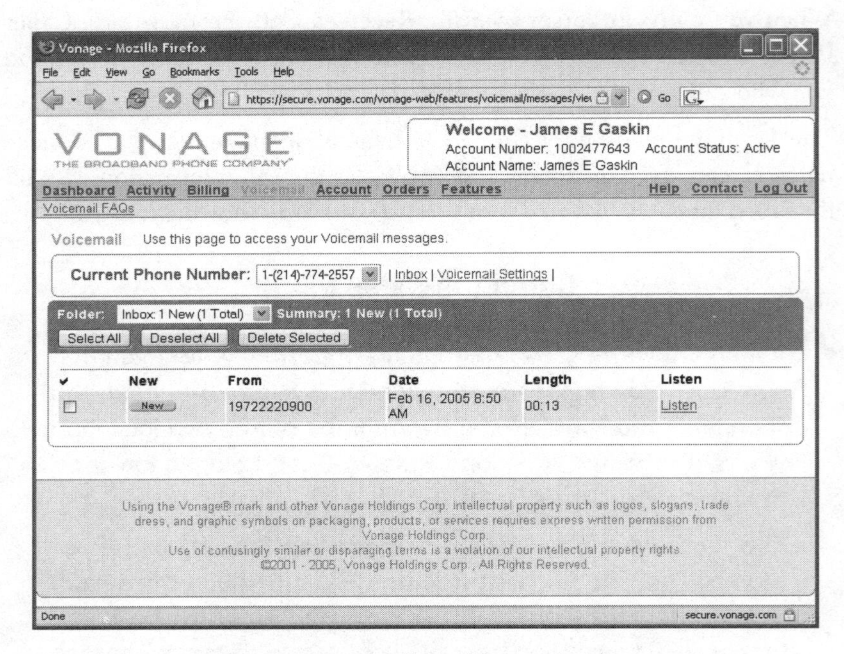

FIGURE 5-7. Check your voicemail from any web browser

Details on your voicemail message include the calling number, the time and date, the actual length, and a chance to listen to the message through your computer. I like this option. Anyone who has a typical answering machine with remote call pick-up probably shares my frustration at getting messages from remote locations. Here, one click and you hear your message.

In addition, I configured Vonage to also send me an email that includes the voicemail message. Less than two minutes after the voicemail message has been left, an email appears with the time, date, calling number, and a sound file containing the voicemail message.

Figure 5-8 shows the Vonage email notification page. You reach this by clicking the "Voicemail Settings" hyperlink in the main voicemail page shown in the previous figure.

Vonage doesn't offer a choice of audio file format, and they use a WAV (Windows Audio Visual) file. This file type means larger file sizes, because the file hasn't been compressed in any way, such as like an MP3 music file. However, this format has the most player options on a Windows computer, so anyone using a Windows operating system will be

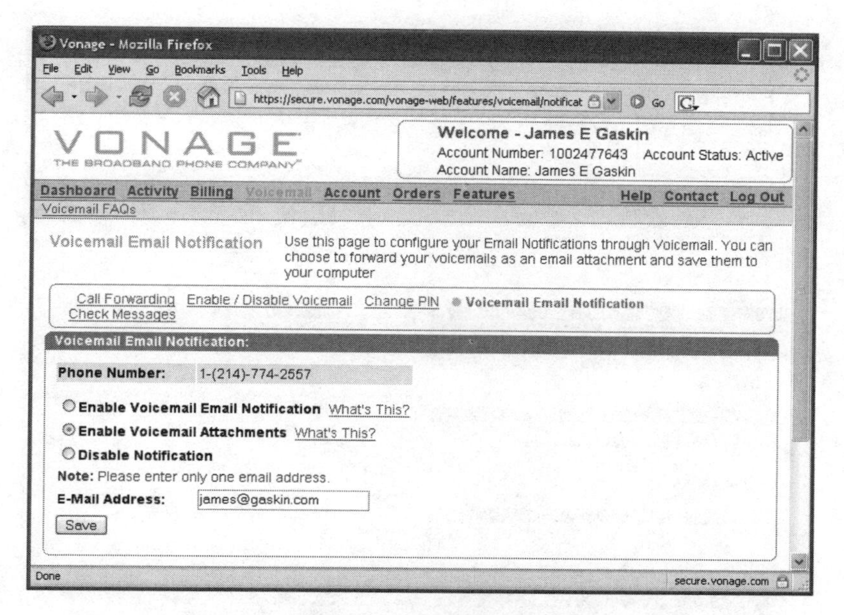

FIGURE 5-8. Configuring voicemail delivery through email

able to play the file when it arrives. Other operating systems also support WAV files without any extra players required, so this is a logical choice.

You can choose to be just notified when you have a voicemail message, rather than have the actual message delivered via email. And, of course, you can decide not to have any type of notification this way, and just find out you have a message when you hear the stuttering dial tone on your Vonage phone the next time you pick it up to make a call.

Managing other features

Figure 5-9 shows the Features page, where you can add or configure the features for your Vonage phone numbers. The first feature to configure, with a red box for attention, is the 911 configuration screen.

When you configure Dialing 911, you see a screen's worth of explanation with two buttons at the bottom: Activate 911 Dialing and Decline 911 Dialing. Vonage strongly encourages you to accept 911 dialing and send your physical location details. So do I.

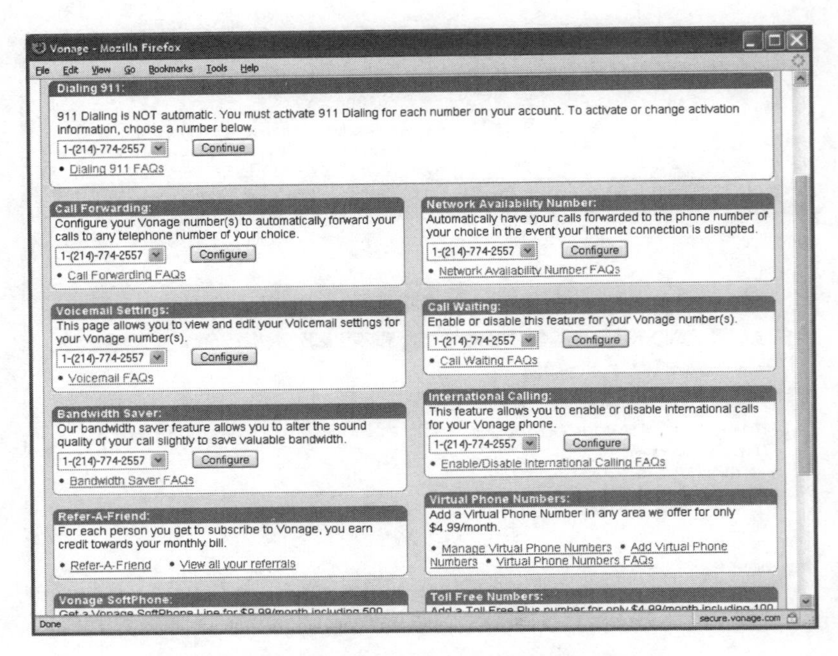

FIGURE 5-9. All your Vonage features gathered in one screen

Let's go through this screen:

Dialing 911

Do this, please. Click Configure and provide your physical address. 911 technical details are in Chapter 7.

Call Forwarding

You can forward calls made to your Vonage line to any other phone line that you wish. Unanswered calls roll to voicemail if they aren't forwarded. You can set the number of seconds an unanswered call will delay before being forwarded (the default is 30 seconds). You can also tell Vonage to ring both phones at once–the one attached to this number, and the one you've listed as the forwarding number.

Network Availability Number

If the Vonage switch can't reach your router because of network or ISP problems on your end, they will forward the call to a non-Vonage number you select. This could be your traditional telephone line if you keep one, your cell phone, or even a neighbor or family member.

Voicemail Settings

Enable or disable voicemail. To reach the screen shown in Figure 5-8, you must go into voicemail and then choose Voicemail Settings.

Call Waiting

Turn call waiting on or off. You must reboot your phone adapter if you make this change. Vonage includes pretty conservative instructions for rebooting. Unplug everything, wait for three minutes, plug in your cable/DSL modem, wait until it completely restarts, plug in your router, wait for it to restart, plug in your telephone adapter (if you have one), wait for it to restart, and reboot your computer. Then check for a dial tone.

Bandwidth Saver

It requires 90 kbps bandwidth upstream to provide full voice quality on a Vonage broadband phone connection (and all other vendor's connections, as well). If you are limited in upstream bandwidth (128 kbps is not an unheard-of limit) and have two phone conversations going at once, the voice quality will degrade in unpleasant ways. By dialing back the amount of bandwidth required for each line, you can lower the voice quality slightly and guarantee your calls won't stutter, flutter, or flake out completely. There is a sliding range setting to use for this configuration. If you have only one Vonage line, you don't have to worry about this at all.

International Calling

If you don't make international calls and don't want anyone else with access to this phone to do so either, use this link to disable international calling and stop worrying.

Refer-A-Friend

This is another example of Vonage's never-stop marketing. I guess since you get service credits for signing up others, you could, maybe, call this a feature.

Virtual Phone Numbers

Want a number in another area code so people there can make a local call and have it ring on your Vonage phone? Here's where you can add a virtual number. The process is very similar to signing up for Vonage initially, except they already have all your financial details and there's nothing to ship to you.

Vonage Softphone

Regular travelers can add a software phone to their laptop for $9.99 per month per line. This softphone is a computer-centric phone and will be explored in detail in the next chapter. But adding another Vonage number for only $9.99 per month is far cheaper than paying phone charges from a hotel or paying roaming charges on your cell phone. Travelers with laptops can save quite a bit of money with this feature. (Of course, they can only call when connected, wirelessly or directly, to broadband.)

Toll Free Numbers

Want to add a toll-free number so anyone in the U.S. and Canada can call you for free? Click the link here and set up that number. You'll pay $4.99 per month and get 100 minutes of inbound calls as part of the package.

A handy page, laid out well, with a link to Help files and more information inside every feature box. Kudos to Vonage for making feature management fairly simple.

Monitoring Costs

Yes, all broadband phone services cost less than traditional telephone lines with comparable services. This doesn't mean you don't want to monitor how much you're paying for the services on your broadband phone service.

The extra-cost features are covered in the Features screens just discussed, so let me focus on the billing and account information pages now. You can get more financial information from Vonage, and make changes as necessary, easier and quicker than any traditional telephone company, including your cell phone provider.

Billing information online

Want to know how much you owe Vonage at any moment? Figure 5-10 shows an account billing page.

My account details will look different than yours because Vonage granted me an account for this book. The 72 cents of charged calls come from one call to London for 24 minutes. Twenty-four minutes to London for less than a dollar? Try that with your traditional telephone line.

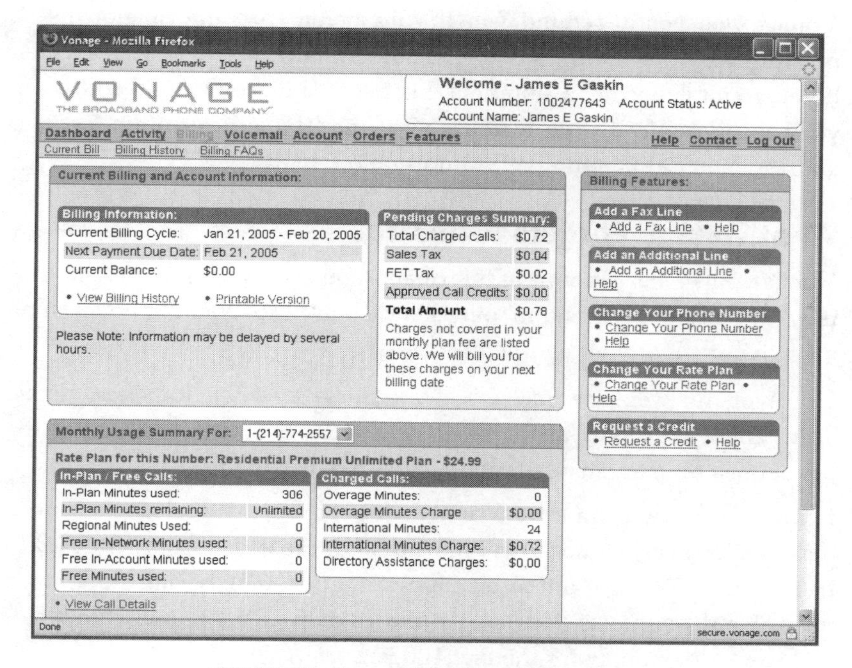

FIGURE 5-10. How much, to the penny, to the day

Want a printout? Click the Printable Version hyperlink in the upper-right box and you'll see a bill formatted for paper. The numbers don't change, but the page arrangement does.

You can drill down on any hyperlinked item, such as View Billing History, and get all your past detail with one click. If you want complete call details, click on the View Call Details hyperlink in the bottom-left corner. You couldn't ever get that information from your traditional telephone company, and many cell phone providers no longer provide this level of detail. Just another advantage of a completely computerized telephone switching network as deployed by Vonage and other broadband phone companies.

Account information

The Account Information page shows all the typical housekeeping details used by any service. This is where you change your address, keep your credit card information current, and change your account password if necessary.

Vonage will gently remind you if your credit card information needs updating for your monthly billing. Your shipping address doesn't matter if you aren't expecting a shipment, but it's nice if the physical address in your account information matches what the 911 service shows for your address. You did fill out the 911 information, right?

What Vonage Forgets to Tell You

Vonage puts a fair amount of information on their web site. In fact, they may put more information out there about broadband phones than most, if not all, of their competitors.

However, no web site answers every single question for every single user. There are items not covered, including Vonage's ranking in the telephone company market and their financial situation.

Being a private company, Vonage doesn't have to release financial details. However, I want to get into some of those details briefly. Not to help or hurt Vonage, but as an illustration of a new business approach (broadband phone services) in the old market of telephone companies that were a monopoly for the first 100 years.

Some of these details Vonage does disclose, but they are less obvious than I would like. Don't take this as an attack on Vonage, because it's not. But there are always overlooked details and information. Here's the place for some of that overlooked (or well-hidden) information.

Technical details they don't mention

First, however, let me get into a few details about the technical side:

1. When Vonage (and their competitors) send you a router that's preconfigured with your phone number, it's also preconfigured to work only with that service. That's the downside of the router being "free"–it's free when connected to Vonage. If you change to another broadband phone service, even one that uses the same types of routers used by Vonage, your free router will become a free paperweight.
2. Interestingly, Vonage Chief Technology Officer (CTO) Louie Mamakos says audiences don't really ask about the technical underpinnings of broadband telephony. He explains it as making telephone calls over broadband rather than over the telephone company and people accept that–except for a few technically minded curious people (like me), who keep asking about little details like jitter and latency and packet loss (described here):

Jitter

A fluctuation or flicker in a transmission, usually caused by mistimed packets arriving early or late.

Latency

The time between when you ask for data and when the data arrives. It often occurs when packets are held up somewhere along their travels, such as the interface between the broadband network and the traditional telephone network.

Packet loss

Broadband phone services work on a "best effort" packet delivery model, so packets that get lost are not resent. The delays introduced by a guaranteed packet delivery protocol would be more noticeable than missing 1/30th of a second of conversation (which is generally the most you need to worry about with the best effort model).

Realistically, the Internet today has so much capacity and has been so well engineered that broadband phone companies start out with an excellent foundation. In fact, Vonage claims that the reason voice quality on their circuits is usually rated 4 out of 5 is because the traditional telephone lines they rely on for connection to non-Vonage clients are rated only 4 out of 5, thus dragging down Vonage.

When you get your broadband phone, regardless of the service provider, I bet you will agree with Vonage. When you call a user on the same broadband system, the voice quality is noticeably better than when you call someone still using a traditional telephone line.

3. Firewalls, especially for homes and small businesses that have been incorrectly installed, can still cause problems. You'll need Vonage technical support to get everything configured properly if you keep an existing router with firewall security that hasn't been adjusted to allow broadband phone software to get through.
4. Standards in the broadband phone world are still developing, and Vonage engineers have written around and extended some of those standards to make their system more scalable. They promise they correctly connect to all other companies using the same standards, but honestly, that sometimes takes a bit of tweaking here and there. Luckily, those issues will cause small problems between Vonage and smaller Internet Telephony installations and will be solved before too many people notice.

5. Caller ID information on calls from other countries is pretty hit and miss. This isn't the fault of Vonage and can't be fixed with another software tweak; it's a problem caused by governments and national phone companies arguing about standards and slowly updating their systems. Caller ID from Canada works pretty well, but don't yell at Vonage if Caller ID from other foreign countries doesn't work.
6. Of the 650 or so employees at Vonage in early 2005, about 150 are technical employees. More than 50 of those are software developers. While this ratio won't impress companies in Silicon Valley, it's a darn good ratio of technical to nontechnical workers for large companies in almost any other location and business.

Business details they don't mention

Here are some of the business details of which you should be aware:

1. There are fees associated with broadband phones, even though they soft-pedal this fact. There are far fewer fees on broadband phone service providers than on traditional and cell phone providers. Vonage mentions this, but the information is easy to overlook.

   Here are the added costs on Vonage's premium $24.99 service:

   - Sales Tax: $1.50
   - Federal Excise Tax: $0.75
   - Regulatory Recovery Fee: $1.50

   Total for the month: $28.74. Compare that to your traditional telephone line's added fees and you're still way ahead with a broadband phone provider.
2. Vonage follows the high risk and high reward path of Internet startups like Amazon.com–lots of venture capital money (about $400 million so far) and lots of marketing dollars spent. Their cost per new customer ranks higher than just about any other broadband phone company.

   While dangerous–and likely to keep Vonage in the red at least through the end of 2006–this is a proven market strategy that has worked in the past. The critical mass of brand name awareness among consumers has been reached: most people believe Vonage is another funny word for broadband phone. Okay, maybe not quite that strong, but the Vonage brand name has become embedded in many consumer's consciousness to such a degree that I believe they have bought themselves a place in the sun as many of their competitors go dark.

I say "bought" on purpose, because they are $400 million deep into pockets of the venture capitalists, and will probably be $500 million before all is said and done. But it's worth it because they will almost assuredly succeed and remain the market leader in phone-centric broadband phone companies for the foreseeable future.

Don't worry if someone, such as a traditional telephone company with deep pockets, runs in and buys them. If that happens, you'll never see a bit of difference in your service. For that matter, customers of any merged or taken-over broadband phone company will continue on, likely never knowing the computers running the broadband phone company switching and routing now sit in a different data center.

3. As we say in Texas, "the big'uns eat the little'uns." That's true in business as well. This could play out two ways: Vonage can absorb some of their competitors and gain their subscribers for less cost than they are acquiring their new customers now because of their high marketing costs; or, Vonage can be the "little'un" and get snapped up by another company that may or may not even be in the telephone business.

Troubleshooting

Continuing with their excellent web content, Vonage has one of the most complete Help sections. There are numerous illustrated installation screens with video-type illustrations showing outstanding detail for each step of each procedure.

Vonage Help Pages

Figure 5-11 shows the main Help screen for Vonage. You reach this screen by clicking the Help link on the top horizontal menu.

The last link on the bottom left of Figure 5-11 is Troubleshooting. Clicking this opens up a list of over 100 topics ranging from general questions to fairly deep technical details. Every router and telephone adapter sold or recommended by Vonage has installation and troubleshooting instructions in this area. You can also reach many of these explanations through the Installation Support links in the top-left corner of the main Help screen.

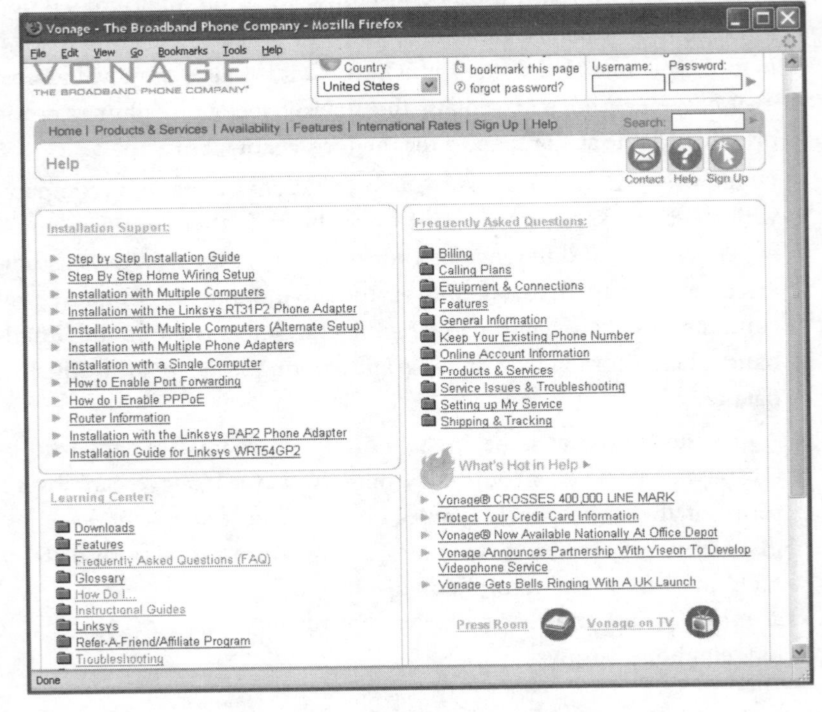

FIGURE 5-11. Lots of help and good explanations

As usual, Vonage marketing gets a large amount of space on this page. The bottom-right quarter of the main Help screen is What's Hot in Help, which is all marketing information such as new customer totals, new retail partnerships, and teasers about their upcoming video service. Oh, yes, there's some generic blather about protecting your credit card information that's a rehash of credit card bill inserts for the last dozen years.

Overactive marketing aside, the Vonage Help screens are unusually well done and full of useful information clearly presented. Again, kudos to Vonage.

Top Troubleshooting Tips

Michael Tribolet, Executive Vice President of Operations for Vonage, says there are three issues that come up the most in Vonage's technical support center. These will apply whether you're on Vonage or another provider:

Wrong plugs

People plug their phones into the wrong telephone port on their router because the print is small on the router or they don't realize there's a difference in the ports. Sometimes things work anyway, but not always. This is no different than plugging your phone into the wrong wall plug in the office or somewhere else you have multiple holes. Check to see that all the cables are connected, and plugged into the proper ports, before calling for help.

Router problems

All the routers Vonage ships are configured properly to let broadband phone traffic through the firewall in a secure manner. But sometimes people changes their firewall configuration and close those ports by accident. In addition, telephone adapters that plug into the routers can't be preconfigured to open the right software ports to allow the broadband voice traffic through the firewall, so users must configure their firewalls themselves.

Relocation

Seven percent of the people in the United States move annually, and many of those are Vonage customers. In fact, if you move regularly, using a broadband phone service as your number makes sense, because you can take your number with you. But what often gets lost or left behind are the instructions for configuring devices after the move.

So you don't have to search, here are the software ports that must be opened in your firewall for your Vonage phone to work:

- SIP (ports 5060 through 5061) using UDP protocol
- NTP (port 123) using UDP protocol
- TFTP (port 69) using UDP protocol
- DNS (port 53) using UDP protocol
- RTP (ports 10,000 through 20,000) using UDP protocol Figure 5-12 shows the Linksys screen used to add these security changes.

Every router is different, so take a look at your manual when setting these ports. The example in Figure 5-12 shows the process, but since this particular router came with Vonage service configured at the factory, changes like this weren't necessary. Please call your broadband phone provider, Vonage or whoever, and ask for help rather than guessing

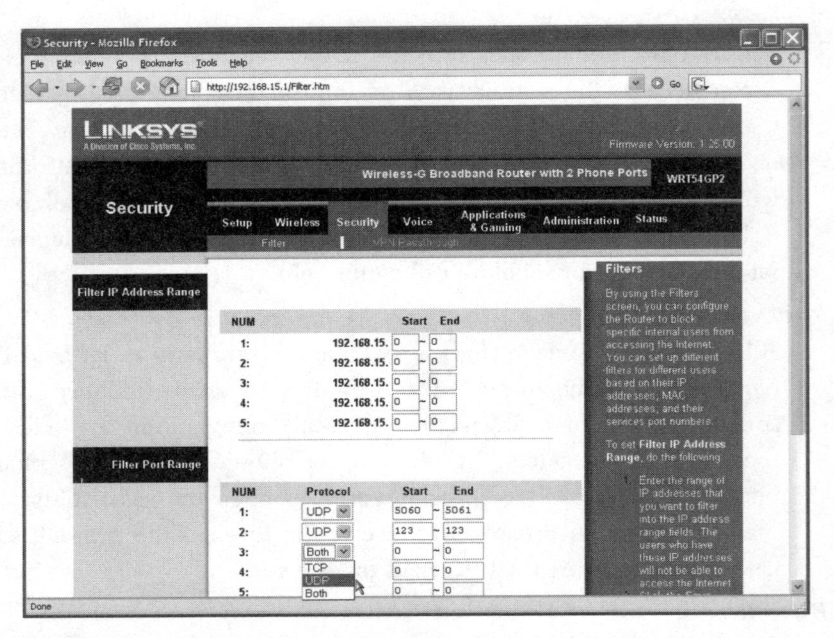

FIGURE 5-12. Configuring a new Linksys router with Vonage support for service access through the firewall

about your firewall. Computers on the Internet need protection, so make sure your firewall is configured properly. You'll have much more fun with your computer if it's safe, I promise.

Reboot

Rebooting fixes more computer problems than anything. The same goes for routers, telephone adapters, and cable/DSL modems. Reboot for better health.

Vonage and many other providers, including many broadband ISPs, have some pretty extreme views on rebooting, more extreme and time-consuming than I usually recommend. But you'll often have to go through the reboot when you call to report a problem, so here goes. (In fact, if you change certain items in the Vonage administration screen, such as Call Waiting, you must reboot.) Here are Vonage's instructions:

1. Turn off your computer.
2. Unplug everything except your computer, meaning your router, telephone adapter if you have one, and your cable/DSL modem.
3. Wait for three minutes.

4. Plug in your cable/DSL modem and wait until it completely restarts and connects to your broadband service provider.
5. Plug in your router and wait for it to restart and connect to your cable/DSL modem and service provider.
6. Plug in your telephone adapter (if you have one) and wait for it to restart and connect to your router.
7. Reboot your computer and wait for it to restart.
8. Pick up your Vonage phone and check for a dial tone.

Personally, I always recommend unplugging things and waiting for one minute. But if that doesn't work, try the three minutes Vonage recommends. Since many people get impatient and turn things on too quickly, maybe telling them three minutes will make them wait at least one minute.

Redial

If you want to sign up for a Vonage broadband phone line, everything you need to know is in this chapter, including screenshots of what you will see as you sign up. If you want to sign up for another company's broadband phone service, you have an excellent idea of what you must provide and pay for the service.

Beyond the initial sign up, you should feel comfortable with the stability of the broadband phone market in general, and with Vonage as the leading player in particular. This market remains dynamic (meaning no one knows what will happen), but Vonage and their competitors are here for the long haul. Sign up without worry, because you'll be able to use your broadband phone from now on without any major problems, no matter what happens in this market or to your broadband phone service provider. If you're company is a "little'un," your service will be transferred, as is, to the "big'un" that ate your provider.

6 SKYPE AND OTHER COMPUTER-CENTRIC SERVICES

Let me be up front: Skype and other computer-centric Internet Telephony providers will not satisfy most traditional telephone users. Skype's ease of use and community drew millions, but they remain tied to their computer and can't get incoming calls from non-Skype users (yet). Skype's competitors are niche products that appeal to a techno-savvy demographic, and will remain so for many years. Skype will not replace traditional phone lines in more than a handful of households.

However, these services now pioneer new types of communication services far beyond what phone-centric providers have to offer. Computer-centric phone applications remain true to the rule that technical trailblazing products never have mass user appeal, but are fun to watch and provide great benefits for a few early adopters (although the early adopters in this case number in the tens of millions). Skype cleverly hides their technology and makes their service fun–something I wish other technology companies could emulate.

Skype's slogan is "Free Internet telephony that just works," and for the most part, they're exactly right. Other softphone applications seem to follow the Skype model but use different telephony standards and business models. But once you have your computer connected to the Internet via a broadband connection, it's easy to make voice calls with Skype. And Skype happily avoids any problems getting through your firewall.

Different software phone services have more problems. It takes some effort to get other softphones working beyond what it takes to get Skype going. You must deal with the same types of firewall port concerns discussed in the previous chapter for the phone-centric providers.

This chapter could be titled "Talk Is Cheapest" because of the pricing structure used by Skype and competitors: free calls anywhere in the world to other users of this service. While you might assume computer-centric telephony provides value only to technology-obsessed youth who are never far from their computer, Skype reports that half of their customers use the service at least part of the time for business.

Skype and Competitors

It may seem that Skype appeared from nowhere and pioneered the idea of a software phone using a peer-to-peer connection model. There was nothing, then boom, there were millions of Skype users. But that's not exactly how it all developed.

Active Internet Telephony pioneers were creating the foundation for years, building on earlier video-conference work, but 1995 may be the year technology gelled enough to get serious (a company named VocalTec released the first Internet Phone on February 15 of that year). The maturing Quality of Service protocol, necessary to keep voice streams traveling over the Internet from becoming garbled, provided an excellent starting place for developing the standards for Internet Telephony.

Acronym Alert

SIP = Session Initiation Protocol

Today, we have Skype on one hand and SIP (Session Initiation Protocol) on the other. Skype development comes from a small group of talented programmers and marketers responsible in part for the popularization of peer-to-peer applications through their first success, the KaZaA file-sharing service. SIP development comes from a larger group of talented programmers distributed throughout the world and coordinated by the oversight group that developed and standardized (among many other things) networking and the Internet. It's a fun competition, and we consumers will be the better for it, although it will be a bit messy for the next couple of years.

The SIP users have one thing the Skype folks lack (at least now, in early 2005, while the *SkypeIn* incoming call service remains in limited beta): the ability for traditional telephone users to call them. Some SIP companies offer virtual numbers that work just like the ones from Vonage and their competitors (discussed in Chapter 5). A SIP phone user, even when they use a softphone and headset, can be called just like any other traditional or cell phone, if they have purchased an optional virtual number.

There are also a few *gateways* from the traditional telephone world to the SIP phone provider networks for the users who don't want to pay for virtual numbers or have little need for them on a regular basis. Traditional telephone users can call these gateways (none are toll free), then dial the special SIP phone number they wish to reach.

Skype's SkypeIn service will provide virtual numbers for Skype users so they can take calls from traditional telephones, but it's still in beta testing as I write this. It will be a paid service, like the virtual numbers for Vonage and the SIP phone providers. No official pricing has been announced, but rumors from the beta say the price should be around $3 to $4 per month for an inbound phone number. The SIP phone world already has its *SIP-In* service, and it is ahead of Skype, at least for the time being.

The SIP World

Technology companies follow trends, and one trend today is for SIP-based telephone developers and services to use "SIP" in their name somewhere. Here are some examples:

- *www.sipphone.com*
- *www.sipsoftware.com*
- *www.sipcenter.com*
- *www.sipforum.org*

There are many other companies in this market without SIP in their Internet address, but they usually have SIP at the top of their web pages and product literature. That makes sense, because this remains a developing market full of companies struggling to make their place and gain name recognition. In fact, the entire SIP world still needs more name recognition.

SIP

Technically, SIP works much like HTTP (HyperText Transfer Protocol), the supporting protocol for the World Wide Web. It's also lightweight, open, flexible, and text-based. Messages include headers (address and routing information) and the body. SIP is an application layer protocol (high up on the network protocol ladder so it can traverse a variety of network types) and controls session creation, modification, and termination.

SIP messages set up calls, create the connection for the call for underlying protocols to handle, and send disconnect instructions when the call is terminated. SIP also works for multimedia conferences and multimedia content distribution, so you will see some interesting things from SIP companies beyond telephone calls in the next two years.

SIPphone

The great thing about SIP and softphones using a SIP foundation is how easy it is to get started and try them for yourself. Take a look at Figure 6-1, the opening web page for SIPphone.

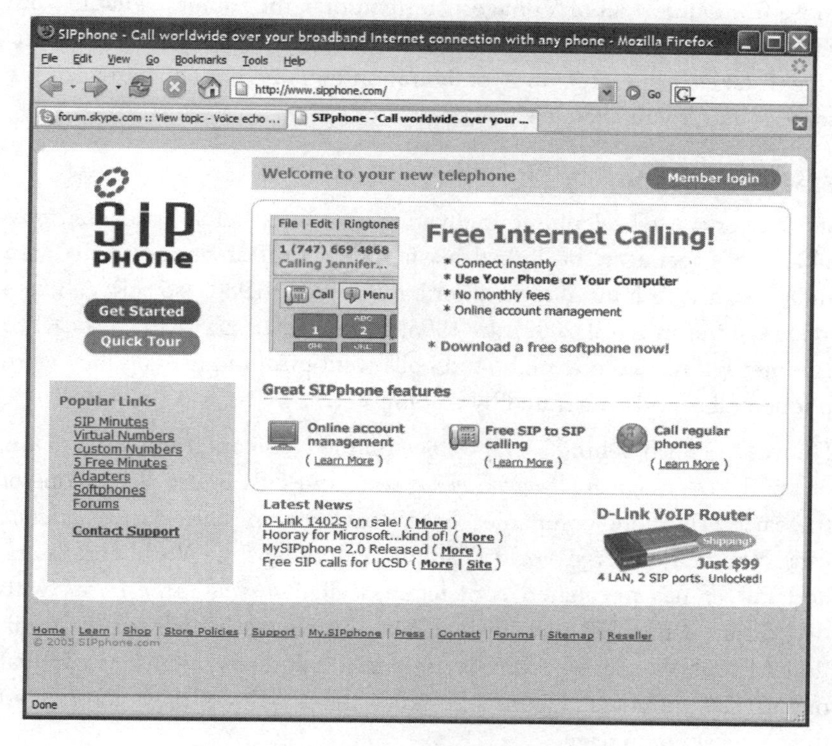

FIGURE 6-1. Download and call from your computer

You can download and install the software and be talking with other SIPphone users in just a few moments. SIPphone also allows you five free minutes of connection time to traditional telephone lines.

Interestingly, SIPphone was started by the man who created the best free music download site, MP3.com (since sold off because of record industry harassment and not nearly so cool today). Michael Robertson also started the popular Linux operating system company called Linspire (*www.linspire.com*), which used to be named Lindows until Microsoft sued them because they thought the name Lindows infringed on the name Windows. And you thought the computer business was boring.

Notice the D-Link router in the bottom-right corner of Figure 6-1. Just like Vonage, many SIP vendors allow you to connect existing analog telephones to your broadband connection. Unlike the Vonage units (and those from almost all of Vonage's competitors), the router available from SIPphone is "unlocked" and will work on any broadband phone service. It takes more setup and configuration initially, but you can keep the same router if you change services.

FreeWorld Dialup

Jeff Pulver started telephone innovation way back before it seemed possible. As a teenager, he linked his family's cordless phone to his ham radio set, giving himself a car phone in the early 1980s (so said *BusinessWeek* Online in April 2003). In 1995, he stumbled across the VocalTec Internet Phone late one night and bells went off. I'm guessing they were phone bell sounds generated by a computer.

He was the push behind VON (Voice on the Net) conferences (*www.von.org*) and *VON* magazine (*www.vonmag.com*), and Pulver also started one of the early SIP phone companies, FreeWorld Dialup (*www.freeworlddialup.com*). You can use FreeWorld Dialup to call other FreeWorld customers, and Pulver has negotiated connections called *peering arrangements* with two dozen other SIP providers. He also includes broadband to traditional telephone gateways so his users in a half-dozen countries can dial out to 800 numbers. Figure 6-2 shows the FWD (as they acronymed themselves) home page.

Pulver organized the VON conference advertised in the banner ad on the lower-left side of the page, and he has gathered together the smartest people in the world of Internet Telephony each year since 1996. It's safe

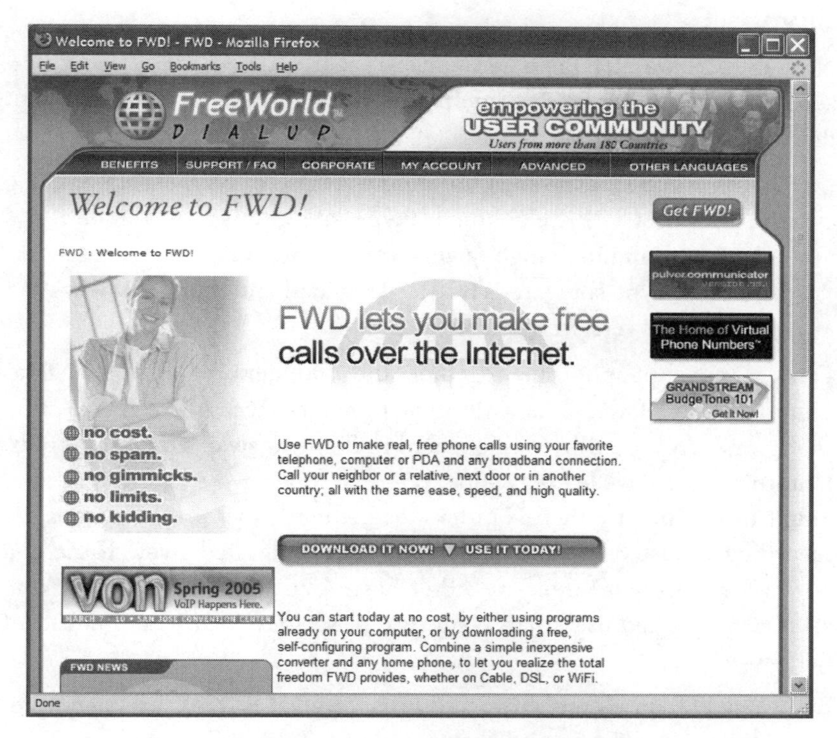

FIGURE 6-2. Free calls from a pioneering service

to say that Jeff Pulver should be named the matchmaker for a huge number of the companies in this market. He also cofounded Vonage, but left early after their initial venture capital investment round of financing.

Yes, there's a peering arrangement with Vonage. The FreeWorld Dialup information even includes the special prefix number you dial from Vonage to reach a FreeWorld dialup user, something Vonage doesn't make easy to find (at least I couldn't find it on Vonage's site).

It's hard to bet against Jeff Pulver and FreeWorld Dialup, and it's hard to bet against SIPphone and Michael Robertson. They've both done interesting things and have records of success. If you know someone using a traditional telephone who wants to call you occasionally, however, SIPphone is a bit farther down that road than FreeWorld.

Skype

Nobody was waiting for a new peer-to-peer phone service using personal computers for phone equipment back on August 29, 2003, but there it was:

> We're proud to announce the public release of Skype Beta and the arrival of P2P Telephony! Skype allows anyone to experience free, unlimited, high-quality voice communication over the Internet. Skype software is free to download and you can always get the latest version by clicking here.

Few, if any, people put together the founders of Skype, Niklas Zennström and Janus Friis, with their previous success, the KaZaA file-sharing network used by millions and millions to swap files of all kinds. Unfortunately, the files they most often swapped were music files they didn't own, since by then Napster was getting heavy pressure from the recording industry to close and many users migrated over. KaZaA, a pure peer-to-peer technology, didn't have a centralized server tracking all their registered users like Napster did, so they kept a step ahead of the lawyers.

But we can't blame Zennström and Friis for what KaZaA has been up to the last few years, because they jumped out. They call Skype their third generation file-sharing network project, and the best one to date. They are right.

Although KaZaA owns the record for downloads at over 370 million, Skype already has a 100 million downloads and counting less than two years after their initial beta release. Skype is also free of all the spyware and other garbage that's given peer-to-peer file-sharing programs (KaZaA to a lesser extent than most others depending on who you talk to) such a bad name the last few years, so download without fear.

Based in Luxembourg, Skype has huge uptake in Europe and Asia but less in the United States. Part of the reason is that long distance call pricing in Europe often staggers the wallet, and most European countries have higher broadband bandwidth than the U.S.

Companies wishing to integrate Skype into their own communication plans can do so, since Skype offers a Software Development Kit. I believe the use of Skype on PDAs with wireless broadband connections makes a great deal of sense for mobile employees who need to talk to other employees. Push one button, connect to another employee, talk, and don't pay a penny for the phone call. Even if one employee is in Seattle and the other is in Miami, not a penny is required. As long as both have access to the Internet, they can talk as much as they want for free.

In early 2005, there were over 75,000,000 Skype software downloads, but many are duplicates as existing users download upgrades. Over 20 million unique usernames were registered by that time. Often, two million members were online and active at the same time.

Stumbling Blocks

Today, getting a software phone and making calls out to other users of the same service is a snap. Even calling out to traditional telephone users works well and costs little. But getting inbound calls from traditional telephone users still includes plenty of hassle.

Getting the Firewall Out of the Way

When using a router supplied by a broadband phone provider such as Vonage, many connection issues are handled for you. The router's security configuration takes into account the attached broadband phones and initializes the firewall security settings properly to allow full phone access.

Using a software phone when you have a typical broadband router with even minimal security makes things more complicated. The problem comes with the way your router hides your internal devices (computers) from hackers on the Internet: the router masks your device's actual IP address. When a request from your computer goes through the router, the Network Address Translation (NAT) portion of the router's security software attaches its own address along with a specific port number.

When return packets (say the web page you asked to view) get back to the router, they include the specific port number given the outgoing packets by the router. The router then connects the outside packets to your device. Outsiders can't see through your router to your internal devices because internal device addresses never go past the router.

Do the NAT Thing

In plain terms, every machine on the outside world thinks it's talking to your router, when it's actually talking to one of the machines on your network. This becomes a problem because your router doesn't know how to answer phone calls. That's why Vonage provides a customized router that knows how to handle incoming phone calls.

This security procedure works great for protecting internal devices from hackers, but also works great at protecting your softphone from callers. These issues will be easier to handle by the end of 2005, but right now they take too much serious time and effort to resolve for me to recommend them wholeheartedly to nontechnical users.

Sneaky Skype

Skype's proprietary application sidesteps most firewall issues.

If you want a broadband phone that is easy to configure and easy for anyone to call, go back a chapter and look at Vonage and the other companies. If you want a free way to call a few other people, each of whom are running the same type of softphone on their computers, SIP or Skype will work great today.

Small businesses (or extravagant home users, I suppose) who put an Internet Telephony server directly on the Internet like a web server don't have these problems. It's only when you try to get through one of the inexpensive home and small office broadband routers with a limited security implementation that frustration grows.

Regulations

Monopoly telephone companies, in the U.S. and every other country, have always been tightly entangled with their governments. That, of course, is how they get and protect their monopolies.

Not the computer-centric phone providers. They are not, according to the U.S. government, phone companies; rather, they are data services companies. In fact, what is now called the Pulver Ruling (yes, that Pulver, as in Jeff) states flatly that these companies and their products may not be regulated like a telephone company at all.

Once you pay your license fees for broadband access to your home or small business (and there are fees, you just don't see them on the back pages of your bill unless you look carefully), what you send over those broadband links is your business. Your Internet Service Provider has some regulatory oversight because they use data communication lines from regulated phone companies, but what you put through your broadband connection is your business.

With a couple of caveats, of course. First, not all countries believe this way. More repressive regimes in Europe and Africa try to control everything, which may involve you if you have family or friends in those countries you wish to sign up for the service. One or two South American countries are making noises like this as well, because they see the revenues for their state telephone monopoly falling. Repressive governments don't care what the people want; they just care about their state monopolies.

Second, we haven't had a court case where a computer-centric phone provider gets caught between their customer's privacy and a district attorney. Not that the services won't give up the names and addresses of users when asked, because they will have to. The problem comes with Skype and their use of encryption.

You'll see more details about this in a moment, but the bottom line is that Skype encrypts all phone connections between each computer (or PDA) running the client software. Great for users who want privacy, but really bad for government officials waving around their court order to wiretap a user.

Wiretapping an encrypted conversation won't satisfy the law enforcement officials involved, and they will get mad at Skype. From headquarters in Luxembourg, the Skype founders will be apologetic but probably not helpful. The local Internet service provider will be helpful, but unable to decrypt the Skype conversation. This mess will make headlines, I promise.

Computer-Centric Phone Features

Computer-centric phones have relatively few features compared to the phone-centric providers in the previous chapter. Why? Because phone features are primarily applied to the receiving party, and computer-centric phones still have trouble receiving calls in many cases.

Honestly, the feature list does lag behind even traditional phones. However, many of the features that traditional telephones and phone-centric vendors seem most proud of don't apply to the computer-centric services. Long distance? Cheap on Vonage, but free on Skype. Caller ID? Skype shows user profiles.

There are some features that attract users, of course. The highest attraction value is the power of community. If your company or a group of your friends use a computer-centric service, you're likely to sign up for that service. The easiest way to get involved with one of these softphone-based services is to sign up in pairs. You and your friend or family member plan to communicate over long distance, and these services offer you a free way to do so.

Money, of course, is the second reason to sign up for Skype or their SIP cousins: they are free to install, free to register, free to use, and free to disconnect when you're tired of them. No shipping fees (at least not for the software phones), no activation fee, and no fees of any kind from state or local governments.

Standard Skype Features

There aren't a lot of features, but the ones here work well. While all the computer-centric phone providers offer free calls to all service members all over the world without charging a penny for long distance, Skype has the largest community of users to call.

Here is a list of Skype's standard features:

- Talk free to other Skype users
- Conference call up to 5 total Skype users
- File transfer
- Chat

That's a short list compared to Vonage. Let's drill down a bit more.

Talking free to other Skype users means connecting to one of over 20 million people worldwide who have registered a Skype username. That's the largest community by far in the broadband phone world. AOL has more potential broadband voice users if you count all their users with the ability to use voice chat, but most AOL users aren't yet on broadband and AOL has just barely started to add broadband phone support. Skype still wins.

Remember how Vonage made a big deal about three-way conference calls? Those are handy, but many times you have a few more than three people who need to get connected to a phone meeting. Skype supports five concurrent users in one conference call (more concurrent users are planned in a future upgrade, but Skype doesn't say how many).

I was surprised that Skype thought it worth the effort to include file transfer in their phone application, and then I remembered the developers also wrote KaZaA. Since they know as much as anyone about file transfers and already wrote the file transfer application, why not throw it in there? Most people don't use this feature, but those who do really love it. Transferring files will impact your bandwidth, however, especially if you're uploading files while trying to converse. Your upstream bandwidth is almost always far less than your downstream bandwidth, so that's where you feel the impact.

Chat? They put instant messaging inside a voice application? While it seems odd, you can instant message separately from your conversation, so it's easier to make fun of coworkers while stuck in one of those painful conference calls. I just wish they came up with a better term than "chat." Chat always meant voice, but they've defined chat as text messaging inside a voice application.

Meeting Tip

It's considered rude to IM (chat) with a buddy on the same boring teleconference, but only if you get caught.

Figure 6-3 shows the Skype home page with their excellent graphic explanation of features. There really aren't any mysteries here, once you get past the "your computer is now your phone company" leap of faith, and one reason people accept this so easily is because of how well Skype presents themselves.

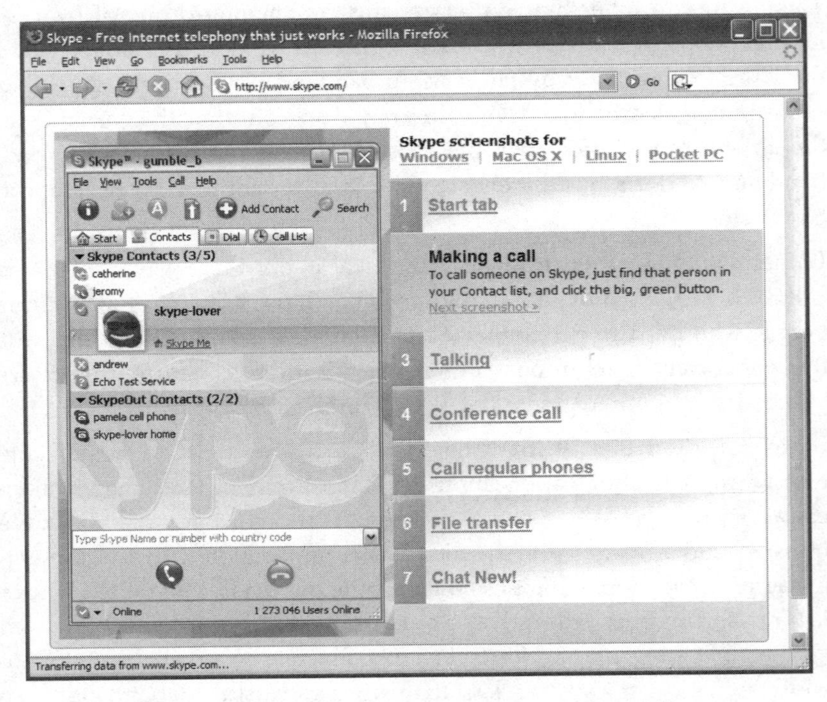

FIGURE 6-3. The second of seven descriptive screenshots at Skype

These screens in the official demonstration bypass the setup and initial configuration part of Skype and show how it works for the millions of users on a daily basis. It's interesting to see what the Skype folks believe is important for the sales pitch.

Notice they put "Making a call" as their second screen, making it the first function after the Skype application home page (you'll see one of those shots later). Their next screen shows a call in progress, then a conference call, then calling "regular phones," which are the traditional telephone lines and cell phones. Four of the seven example pages they show are about talking, which makes sense because Skype is a voice application with some extras, not some other type of application with voice grafted on. Skype was built from the beginning to change the voice connection landscape, and they've succeeded.

Skype works on a decentralized model, meaning there are no big telephone switches and controlling computers in regional data centers, such as those that power Vonage. The only centralized Skype services at the beginning were the login servers, which also show which other Skype users are online, and the switches controlling SkypeOut calls to traditional telephone networks around the world. Adding Voicemail and SkypeIn (both still in beta as I write this) require more centralized resources, and Skype data centers are growing as you read this.

Voicemail applications for your computer from third parties are under development, but Skype won't have their own until they release Skype Voicemail officially. Skype's beta program indicates Voicemail will be a subscription service costing 5 euros per 3 months (told you they are European). So while voicemail is a standard feature for every phone-centric broadband phone provider, it is not a standard feature for Skype (but is for SIP-based competitors).

Call forwarding isn't included because only a fraction of the total Skype users pay for the optional SkypeOut service. Rather than forwarding calls from, say, your computer to your laptop if you're at the stereotypical coffee shop, Skype rings every device you are logged in on when you get an incoming call. This way, they need only track your name and password one time, even if you're currently "on" two devices at once. It's not call forwarding, but it works out just as well if you're waiting for a Skype call while you're away from your primary computer.

There is no provision, now or in the future plans Skype expressed to me, for 911 support. Skype cannot be the single telephone link for people at this time, because their SkypeIn service isn't yet released, making it impossible for traditional telephone users to call Skype users. People using Skype must still have another telephone and should rely on that phone for 911 service.

Advanced Skype Features

There are no "advanced" features as far as Skype claims, but there is currently one optional ($$) feature, and some of the technology they integrate I consider advanced. But let's start with the official optional cost product.

SkypeOut is the connection between Skype and the traditional telephone network. Currently, this is a one-way mirror type of gateway, since Skype users can call out to traditional telephone users but can't receive return calls from those traditionalists.

There are two pricing tiers: a global rate for the two dozen most popular destinations, and individual country rates for the rest of the world (and Alaska and Hawaii). Since the Skype folks have a decent sense of humor and do a good job explaining themselves, let me quote them describing their most popular destinations:

> Argentina (Buenos Aires), Australia, Austria, Belgium, Canada, Canada (mobiles), Chile, Denmark, France, Germany, Ireland, Italy, Mexico (Mexico City, Monterrey), Netherlands, New Zealand, Norway, Portugal, Russia (Moscow, St. Petersburg), Spain, Sweden, United Kingdom, United States (except Alaska and Hawaii), United States (mobiles) and last but not least: Vatican.

Notice that not all of a country is covered just because some of a country is covered. Connections from a broadband phone company to the traditional telephone network have to be negotiated for each location. Countries with less-developed telephone networks can't take a single input from Skype and deliver those calls all over their country because they don't have the technical foundation to support that much extra traffic.

Skype does not include Alaska and Hawaii in the standard rate, which is about $.02 per minute. Being European, Skype lists the pricing first as 1.7 euro cents per minute, and 1,1 pence per minute (Skype has an office in London as well). Two cents a minute for long distance across the world matches the best pricing from Vonage and other phone-centric providers.

Skype installation and firewall handling beats the SIP family phones by a considerable margin. Although installation shouldn't be called an advanced feature, handling security details with the firewall can. So I consider the rest of the Skype advanced feature set as described in the following sections.

PDA support

The idea of cell phones and PDA functions mashed together thrills many people tired of filling two jacket pockets with electronics. One approach is to add PDA functions to a cell phone, such as the Treo and SmartPhone. Another approach is to add phone functions to a PDA. Skype, and a variety of SIP products, decided to go the "make your PDA talk" direction.

Owners of a Pocket PC running Microsoft Pocket PC 2003 operating system with Wi-Fi support and at least a 400MHz processor can download the Skype software and turn their PDA into a phone booth, albeit a very small one. While Skype isn't the only service with PDA software (Xten makes the softphone sold by Vonage and has a Pocket PC version at *www.xten.com*, and a beta service called Stanaphone also has software at *http://stanaphone.com*), they are the only service with 20 million active subscribers.

Pocket PC Skype users can participate in conferences but can't initiate conference calls. You must add your contact list manually to the Pocket PC at this time, but a utility for transferring the contact list will be available before too long.

You can use a Pocket PC for phone calls without a headset, but that would be rude unless you can talk into it like a regular handset (yelling into it like a walkie-talkie makes people laugh at you). Get one of the many headsets or earbuds for your Pocket PC and make both parties in the conversation happier.

Presence

The Pocket PC version of Skype adds another advanced feature, although it's not so much a feature as a communications philosophy. *Presence* is the technical name for services that track you wherever you go. I believe Skype adds a new look at presence, tracking individuals where they are most often during the workday–at their computer–by watching their keyboard activity and providing information to others looking for you.

Figure 6-4 shows the presence options Skype provides. You can let these change based on your system activity or put up the Do Not Disturb sign yourself.

You can see in the bottom-left corner of my Skype window that I am currently online (at least when this screenshot was grabbed). There are seven status options available to indicate, or hide, your presence:

- Offline
- Online
- Skype Me
- Away

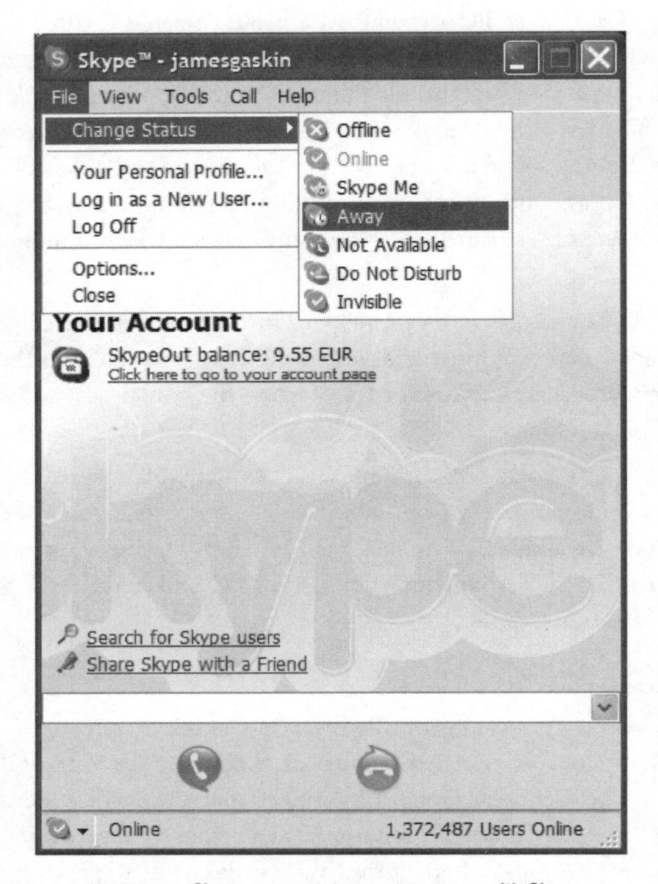

FIGURE 6-4. Change your status, or presence, with Skype

- Not Available
- Do Not Disturb
- Invisible

Online and Offline are pretty clear. Skype Me is the welcome mat laid out for the world to see (literally, with Skype, because people from all over the world can see your status).

Away means you are logged in to Skype, but haven't touched your computer for five minutes. Not Available comes after Away, and kicks in after 20 minutes of inactivity (those are the default settings).

You decide whether to add Do Not Disturb, which, just like hanging the sign on the hotel room door, says you're in but don't want anyone to know that. Invisible means you are online, using Skype, but everyone else sees your status as offline.

This combination of user-set status announcements and those that kick in on timing based on your computer activity is unique. Skype even tells people checking your status in their contact list how long you've been away from the keyboard. No other phone system I know, broadband or traditional, offers so many status options.

Presence means more than just a Do Not Disturb tag on the metaphorical door. Skype rings all devices where a user is logged in at the same time. If you take your Pocket PC out of the syncing cradle and walk to the coffee shop, and someone calls through Skype, your computer and your PDA will ring. You get the message, and the caller is none the wiser about where you are. All they know is that they wanted to talk, and you answered.

Although not strictly presence, Skype's support of Instant Messaging and file transfer adds to the advanced nature of communication. People working together often need to exchange files. Email attachments work, of course, but if you're speaking with someone over Skype and they need a file, you don't have to switch to another application, look up the person's email address, find the file using the email program's explore function, attach, and send.

Clever Techs

My friend Phil says many tech support folks will send usernames and passwords over Skype's instant messaging but never through email, because the Skype connection is encrypted while email is not.

Contrast that process with Skype file transfer–drag the file to the online contact within Skype and drop it there. The recipient then has a chance to use the Save As dialog box to store the incoming file wherever they wish. You can drag multiple files at once, and file sizes, no matter how large, aren't a problem. Files, just like voice streams, are encrypted from end to end when traveling through Skype. But they move slooooowly.

Those interested in more unusual Skype applications will enjoy Chapter 8 (hint, hint).

Future Skype Features

Skype needs more features. Lean and mean is one thing, missing details like voicemail and incoming calls from traditional telephones are a major problem.

Voicemail

How long after you got your first phone line did you get an answering machine? Maybe a day, maybe two?

Skype left this huge hole by not having voicemail (which they call SkypeVM or Skype Voicemail depending on the news source) from the beginning through the spring of 2005 (nearly two years). They haven't yet released SkypeVM, but there was beta software on the Skype site (it's hard to find but there) in early 2005. Perhaps by the time you read this, they'll have SkypeVM up and running. I bet they will, because now Voicemail shows up in the context menu when you right-click some contact names, and they have released subscription pricing.

Voicemail implies a centralized service because that's what the traditional telephone companies sold from the start. After all, if you buy an answering machine, even from the phone company, they get money once. If you subscribe to voicemail, the telephone companies get money every month. Telephone companies like getting your money every month.

I didn't think Skype would absorb the cost of centralized servers to grab, hold, and replay voice messages for free, and they didn't (5 euros per three-month subscription according to the pre-release information on the Skype web site). Whether Skype will also roll it into a more expansive (and expensive) offering remains unknown.

The beta information makes a point to mention that Voicemail moves over the Internet while encrypted, just like the rest of Skype's traffic. Files on your local computer are decrypted automatically when downloaded.

Enter SkypePlus, another future product announced with their typical startling lack of clarity. Hints so far promise the voicemail product along with better conference call–handling capabilities. When? Soon, they say, soon.

Some third parties have developed voicemail, or more accurately personal computer answering machines, for Skype. One of the most popular is called SAM for Skype Answering Machine (a nice personalization). So far, it's free for the downloading at *www.freewebs.com/skypeansweringmachine/index.htm.* If the product takes off (as it may if people get tired of waiting for the official Skype product), search the Web for "Skype Answering Machine" because the developer will have to get a real web site–hosting company without host-advertising banners. It will probably also stop being free.

Speaking of paid Skype enhancements, one of note is the Skype Forwarder, which includes an answering machine and is available at *www.twilightutilities.com/SkypeForwarder.html.* The current price is $19.95, but there's a free demo offer on the site. If you want to forward calls, you'll need a modem in your computer that's able to dial out so you can make that connection.

Call Antiques Roadshow

Modems? Remember those? Since broadband became so popular, modems have seemingly disappeared. Check your garage and junk drawer to find an old one.

Personally, I believe this delay in voicemail support, and the lack of third-party solutions for a program with 100 million or so downloads points out a problem with Skype's proprietary development. On the other hand, no one's downloaded 100 million of any competitive peer-to-peer broadband phone applications, so it's hard to gripe too much about Skype's philosophy.

SkypeIn

The complement to SkypeOut, SkypeIn, is under development and active testing. No formal announcement has been made, although Skype officials have said they hope to get SkypeIn rolling out to the world by summer 2005.

Skype stuck to their peer-to-peer roots with their SkypeOut service, relying on smaller gateways to traditional telephone networks rather than buying centralized telephone switches as have the other broadband phone companies. This made it possible to expand SkypeOut more quickly, and certainly more cheaply, than some of their competitors facing large capital expenditures.

However, the other side of that decision is trouble handling incoming calls for the SkypeIn service. Telephone companies expect to interconnect with other telephone companies at big centralized switches, and Skype doesn't have those. Making SkypeIn a success will almost certainly mean Skype must change some of their business model and start paying big bucks for centralized switches like other broadband phone providers.

They may be clever enough to wiggle around that requirement, and if so, more power to them. But no one else has managed this trick. But no one else built their own peer-to-peer phone application from scratch and attracted 20 million registered users, either.

Skype for Business

If half of current Skype users feel they've used Skype for business, even if what they consider business use is what you and I may consider meager, that's a serious clue for the Skype executives. And they are, it appears, making things happen in their own way.

By summer 2005, Skype for Business should be a packaged product available for an (as yet) unknown price per month. Again, these product features require more investment and centralization than Skype has put into place early on.

The Skype for Business features disclosed by a Skype executive speaking at an Internet Telephony conference include:

- Account management gathering multiple Skype accounts into a single business account
- Web-based management interface for controlling and administering those accounts
- More control over which clients ring with a new call (like a PBX searching for available phones defined in what's called a *hunt group*–a group of telephone lines configured so that calls automatically roll over to the next available phone line)

Skype for Business must rely on SkypeIn, which may be causing the delay in rolling this out as a new service. Businesses also demand control over details foreign to Skype currently, such as directories, authentication, calling restrictions, call logging, call tracing, and connection to internal voicemail systems. Whether Skype can include a full suite of business services this summer will be interesting to watch.

Services Comparison

Skype stands to one side, 20 million registered users and 100 millions downloads strong. Many times the Skype status bar lists over two million current online users.

On the other side stand the SIP providers with tens of thousands of users, no way to track them to see if how many are online at once, and no service with over a million downloads yet. Not active or registered users, but total potential downloads doesn't come to even half a million for any service I can find.

Doesn't seem fair, does it? But while it appears Skype has run roughshod over the computer-centric phone world, one little detail sticks annoyingly up. Standards.

Computer-centric features

SIP is a developing standard blessed by the main technical groups overseeing Internet development and future direction. Skype stands alone against a world of standards groups and coordinated development. It has come down to Skype versus the world for computer-centric broadband phone providers. Take a look at the comparison in Table 6-1.

TABLE 6-1. Computer-centric provider comparison

Feature	Skype	SIPphone	FreeWorldDialup
Dialing out	Yes	Yes	Yes
Accept incoming calls	Yes (beta now but soon)	Yes	Yes
Voicemail	Yes (beta now but soon)	Yes	Yes
Instant Messaging	Yes	Yes	No
File Transfer	Yes	No	No
Easy dialing links to other broadband phone networks	No	No	Yes
Search and display other subscribers	Yes	Yes	Yes
Linux and Mac support	Yes	Linux	Yes
PDA support	Pocket PC	No	Windows CE
Market acceptance	Huge	Minimal	Minimal

Let's drill down a bit more into each of these comparison features:

Dialing out

Each service does this, and each service charges money.

Accept incoming calls

Skype lags now, but SkypeIn will catch them up.

Voicemail

Skype Voicemail will once again catch Skype up to the level of the other services.

Instant Messaging

Skype builds IM into their basic client, and SIPphone's Gaim (the open source Gaim modeled after AOL's original AOL IM that supports numerous instant message services including AIM, MSN, Yahoo, and Jabber) software phone includes IM as well. Gaim came out of Linux operating systems first, but the Windows version is now available.

File Transfer

Skype had a head start with the KaZaA background, but the no other company has stepped up to offer file transfer.

Links to other broadband phone networks

FreeWorld Dialup pushes peering relationships between competitive broadband phone services because a rising tide raises all boats. You can call from FWD to over three dozen SIP-based services by dialing a few extra numbers.

Search and list other subscribers

Searching is easiest with Skype, but not by much. Every service allows various search methods, such as location and name.

Linux and Macintosh support

Skype and FreeWorld Dialup cover Windows, Linux, and Macintosh. SIPphone has yet to cover Mac, and supports only Linspire Linux.

PDA support

Skype supports Pocket PC, while FreeWorld Dialup provides a softphone for Windows CE.

Market acceptance

Skype leads the field by far, with more active users online than the other services have total downloads (combined).

Is there one feature you really need that only one service has? If so, your choice is simple and you should sign up for the feature you need. If not, you'll probably sign up for Skype and join the company of 20 million other registered users.

Compared to Vonage

Look at the type of features computer-centric systems leave off; I discussed them at length in Chapter 5:

- Transferring your current phone number to Skype or other computer-centric softphone service provider
- 911
- Caller ID
- Call Forwarding
- Call Waiting
- Assign rings to specific calling phone numbers
- 311 for local information
- Fax line support

On the other hand, none of the computer-centric phone services demand a $15 or $25 monthly payment just to keep the service active.

One feature, the ability to use your existing phone, might seem to be sneaking in under the wire as a regular computer-centric feature because this capability is supported by the D-Link router now sold by SIPphone. But there are also fairly inexpensive (less than $50) converters that link an analog phone to a computer's USB port. Several phone models that plug into computers look like regular phones, and some are cheaper than the converters.

It's impossible to recommend that someone rely on a computer-centric phone as their only "regular" telephone. Many people rely on their cell phone and never subscribe for a traditional telephone line from their local phone company, but that's a better option today than trusting a computer-centric service.

Passion

Skype wins this category hands down. One goal of Skype founders was to recreate the buzz and "viral marketing" of KaZaA, where a program spreads like a runny nose in an overcrowded preschool. They achieved that goal. Skype users aren't quite as fanatical as early Macintosh users, but they love Skype capabilities and Skype community.

The Skype forums are full of people thrilled with Skype. Australian Rotarian groups are signing up en masse. Love blooms, photos are file transferred, meetings happen. Conference calls full of native speakers and those trying to learn a new language create new bonds and better vocabularies.

When were you so thrilled with any telephonic device that you made a movie to help convince your family and friends to switch to your service? Pim in the Netherlands did, and in his Dutch hip-hop lyrics, I caught the phrase "free long distance" several times (*www.webkeutels.com/video/skype.WMV*).

This passion, above and beyond all technical advantages or service shortcomings of competitors, makes me believe Skype will win out over the SIP-based services for the consumer market. Twenty million registered users sounds like a tipping point for continued growth to me, and 100,000 new downloads per day means that registered user total continues to climb.

How to Sign Up

Signing up for Skype is a snap. Go to *www.skype.com* and click on the large green banner in the upper-right-hand corner. Figure 6-5 shows the top part of the Skype home page.

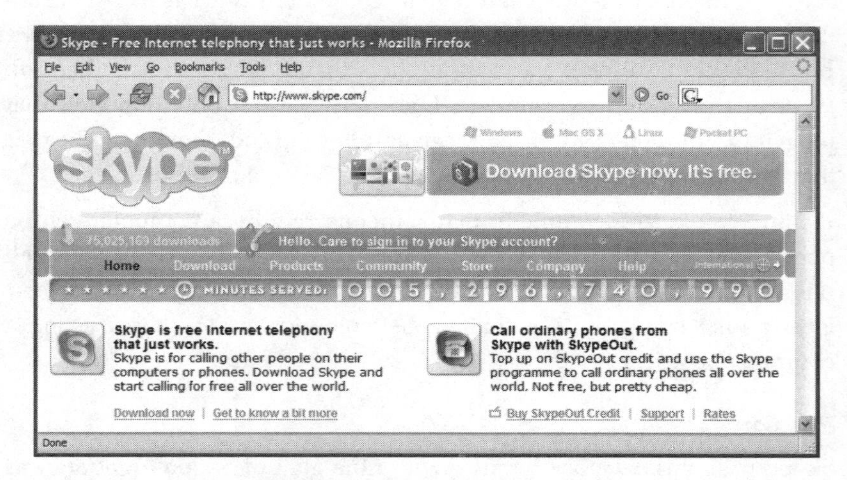

FIGURE 6-5. Easy to download, easy to install

You may not be able to read the small print above the banner, but Skype supports four client platforms:

- Windows
- Macintosh
- Linux
- Pocket PC

Part of the Skype philosophy seeks to make product use so simple that everyone can do it. Their community forum members often brag about signing up parents and grandparents who can barely turn on their computer but become Skype addicts themselves. Figure 6-6 illustrates that approach as Skype provides installation instruction and system requirements on the download page.

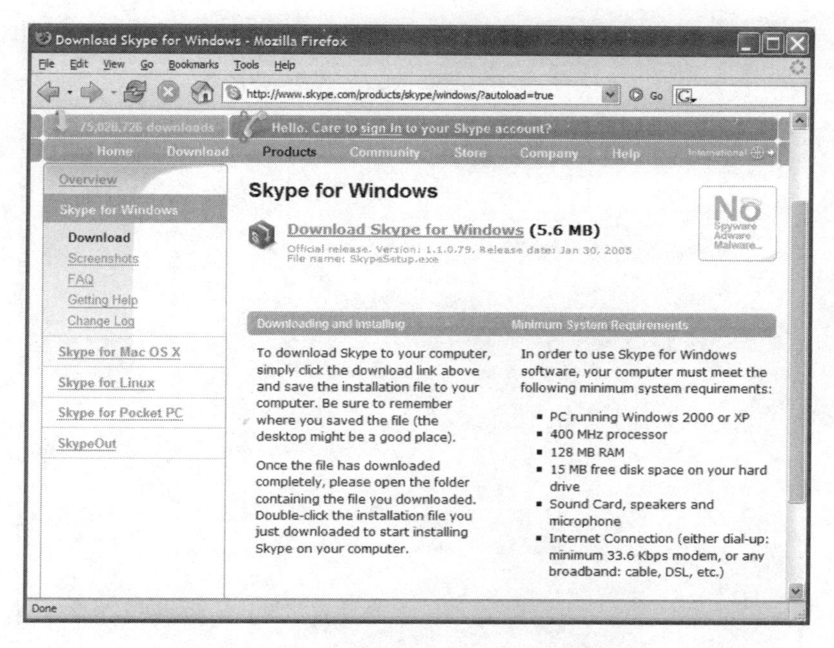

FIGURE 6-6. Clear, jargon-free instructions relax new users

If you don't want Windows software, look to the vertical menu down the left side: all other client software downloads can be started with one click. Skype names the installation file *SkypeSetup.exe* so there's no confusion about which *Setup.exe* file on the computer (many people have a bunch of them) to run.

When the *SkypeSetup.exe* file runs, it will go through the normal Windows installation routine. You must choose your language (Skype supports many), accept the license agreement, and choose the program installation location on your hard disk (take the defaults). You decide during installation if you want Skype icons on the desktop and/or the quick launch bar, whether Skype should start each time Windows starts, and if you want to launch Skype immediately after installation.

Before calling, you must create your Skype username, password, and provide your email address. Figure 6-7 shows this screen.

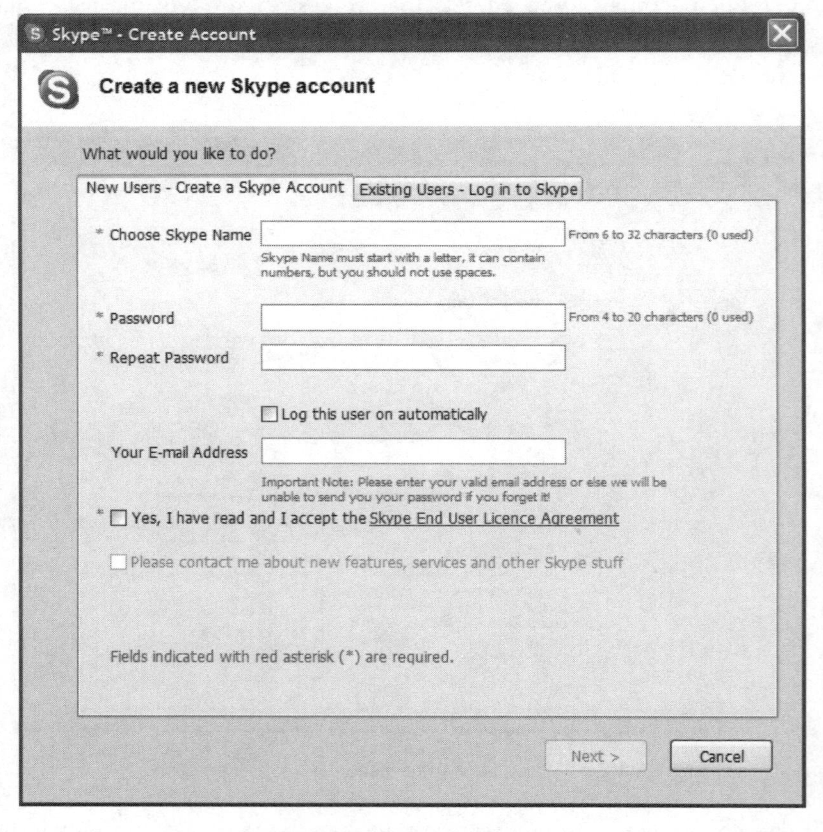

FIGURE 6-7. New account set-up

Your Skype name must be unique and between 6 and 32 characters long with no spaces. Few people will actually have to type your name because your friends will save your name as a contact and just click to call. Your password must be between 4 and 20 characters, again with no spaces allowed.

Email addresses aren't mandatory, but useful to have listed. Skype swears they don't share user information, and I've seen nothing to contradict that statement. Once you provide your email address, the bottom checkbox becomes active and asks if you want to be notified about new features. Say yes.

You have to repeat your assurance that you read the End User License Agreement. You can't use the service if you disagree.

You have a chance to log yourself on automatically. Don't do this if others have access to your computer, because another person can easily impersonate you.

After your name registers as unique with the Skype user directory, you can start making calls. You may want to fill out some personal information before starting, though, so you'll appear more credible to other users and make it easy for them to find you through the search utility. Figure 6-8 shows the personal information form.

Skype not only does not sell your information, they claim they don't store it (in the current version, at least). Perhaps they mean they don't store it anywhere outside their user authentication and login database where they track the number of users online (logged in). Currently, only other users who are online during your search are found, but Skype hints they are adding more of a directory service in an upcoming version.

Details are tracked per client, so if you want different personal information to display when on a laptop or PDA as opposed to when you're on your desktop system, just fill out those profiles differently.

Fill in as much information as you feel comfortable advertising to the world. Sad to say, but there are a few jerks on Skype who search for particular types to harass. You can block users, but you know will know whom to block only after you've been through an unpleasant situation with them.

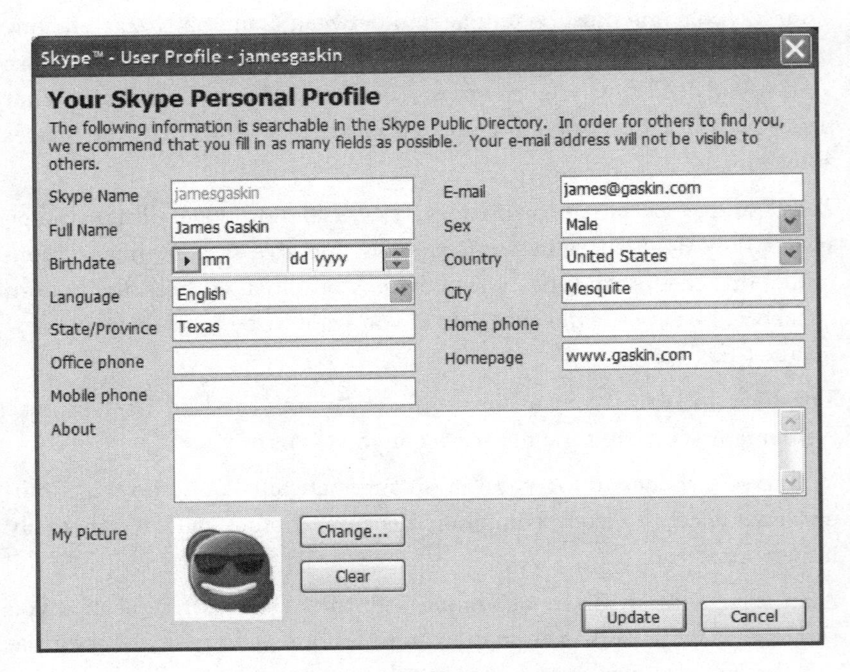

FIGURE 6-8. Optional, but helpful, personal listing

The Skype icon in the bottom right offers some flavor to your look. Skype includes nearly 40 different icon people, and you can add your own image. It's simple:

- Click the Change button beside the picture.
- Click the Add button in the My Picture window that opens.
- Choose your new picture using the file browser that opens.

Click the Set button to save your picture as your current picture. Figure 6-9 shows my new picture. Do you like it?

Okay, that's not me, that's our Sheltie, Hunter. But people always say, "Isn't he cute" when they meet him, and nobody says that about me anymore, so I thought his picture would make a better impression.

You're limited to images with either a JPG or BMP extension. Skype also offers avatars on their web site now and then, but they tend to play hide-and-seek with them.

Once you have your personal profile finished and your picture chosen (optional, but fun), you can start calling anywhere in the world to other Skype users.

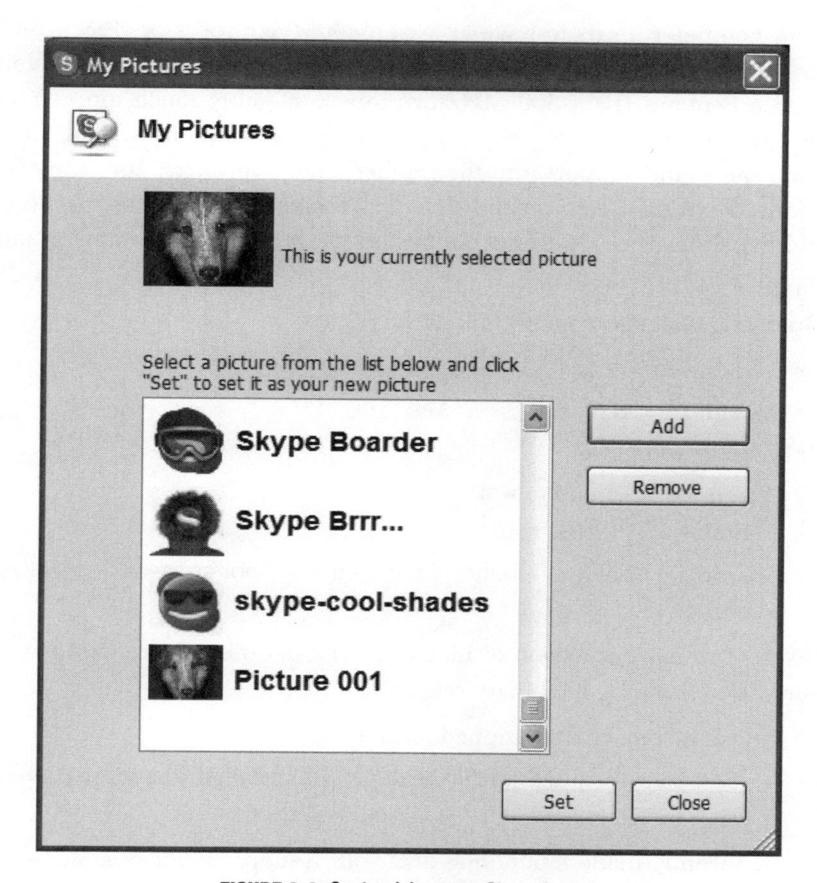

FIGURE 6-9. Customizing your Skype image

The process for SIPphone and FreeWorld Dialup works much like this, although those services are more business oriented, so they register you and send you verification via email. Skype is more open and more consumer-friendly and lets you make calls right away. Those other systems don't let you appear to the world as a dog, either.

Requirements

What equipment do you need in order to sign up for Skype? You need a computer, a connection to the Internet, and something to talk in to and hear out of. You don't even need a broadband connection, because Skype works (barely) over a dial-up connection.

Your computer needs to have at least a 400MHz processor, 128 MB of RAM, and between 10 MB and 20 MB of disk space. Realistically, you need a Pentium III of 500 MHz or more, preferably much more. You also need sound support (they say a sound card, but many systems now build the sound support into the motherboard), speakers, and a microphone. I strongly recommend you skip the speakers and microphone and get either a headset or a phone that plugs into your computer and supports Skype software.

Operating systems supported include:

- Windows 2000 and XP (Home and Pro)
- Macintosh OS X
- SuSE 9 and newer
- Mandrake 10.1 and newer
- Fedora Core 3 (Red Hat)
- Xandros, MEPIS, Ubuntu, and other Debian-based operating systems

If you're running a version of Linux that doesn't meet the requirements, you can try running the binary releases:

- Dynamic binary as a bzipped tar file
- Static binary bzipped tar file with Qt 3.2 installed (they also have a version for download with Qt 3.2 compiled in)

Skype recently made a bundling deal with Xandros to include Skype in their low-cost hardware and software systems. Xandros provides these to a variety of retail outlets, including Wal-Mart (online through *www.walmart.com*). The package will include vouchers for up to 120 minutes of SkypeOut.

Bottom line? You don't need to much of a computer or Internet connection to run Skype. The requirements for SIP-based systems are comparable. Notice you don't even need a credit card or billing address for Skype-to-Skype calls.

How Much?

Nothing.

Skype doesn't cost a penny to download and use to call other Skype users. Neither SIPphone or FreeWorld Dialup asks for a credit card when you download their software and install it on your computer. They also offer free calls to other service members.

SkypeOut minutes cost about $.02 depending on the euro exchange rate and are bought in blocks of 10 euros (about $13.30 in early 2005). Rates to some countries are more expensive, including Alaska and Hawaii (considered long distance while the rest of the U.S. states fall into the lower bulk rate of two cents).

SIP minutes from SIPphone also use a flat rate of two cents for calls to any location in the U.S. Check their rate sheet carefully if you have specific countries you wish to call, because the pricing varies wildly and you could be unpleasantly surprised.

FreeWorld Dialup refers users wishing to call traditional telephones to sign up with a third party for access. They recommend IConnectHere at *www.iconnecthere.com* That's a pretty good recommendation, since IConnectHere offers U.S. rates at 1.1 cents per minute. Their international rates are much more in line with other broadband phone services and consistent between countries than those from SIPphone.

Decision Checklist for New Users

First, you have to decide if you want to use a computer-centric broadband phone service. The dynamics vary considerably from normal telephone use and from the phone-centric providers discussed in the previous chapter.

Computer-centric broadband phone services primarily benefit groups calling each other over the service. Many Skype users are invited by earlier Skype users, a process Skype makes easy by adding invitations to their web site. SIP-based broadband phone providers rely on business users to define their membership group.

The first few decision checkpoint items in the previous chapter don't apply here at all. Go through the following table. If you answer such questions as "I need to keep my traditional telephone line" and "I need to keep my same phone number," affirmatively, it means you need to stick with phone-centric providers.

This won't take long, I promise.

Criteria very important	Somewhat important	Not important
The majority of the people I want to talk to are on the same service.	❑	❑
My contacts in foreign countries also have Internet access.	❑	❑
File transfer will be helpful.	❑	❑
I like to use IM as well as phone calls.	❑	❑
I need to conference more than two other people.	❑	❑

The first question does the heavy lifting: if most of your contacts are on one service, you join that service. Since it's free, signing up for a computer-centric service that turns out to be more pain than pleasure doesn't cost a thing. You can even delete the software from your computer and reclaim the bits of your hard disk it used.

The second question will help you decide if you want a phone-centric service or a computer-centric one. If all your foreign contacts have Internet access, then you can talk for free regardless of which computer-centric service you share. If your contacts don't have Internet access, but only a traditional telephone, you still may benefit from Skype or a competitor.

Do you call them most of the time? If so, you can use SkypeOut or the equivalent and pay just the few cents per minute long distance charge but nothing else. You get inexpensive long distance without any monthly fees. You get inexpensive long distance with Vonage and competitors, but there's always a monthly fee as well.

If your foreign friends and family call you as often as you call them, the Skype option won't work today. Go ahead and sign up for a phone-centric service from the previous chapter or one of the SIP phone options. Or wait until Skype officially launches SkypeIn, because their hinted pricing will be quite competitive.

All other things being equal–except that you need to conference call with more than two other people–sign up for Skype. The same goes for file transfer; Skype's the only action in town right now.

Using Skype

To make a call, highlight a contact name in your Contact list, click the green telephone button at the bottom of the Skype application, and you'll hear the phone ring, just like a "regular" phone. The person on the other end of the line will pick up their receiver (or headset) and say, "hello" in whatever language they prefer. You're talking via Skype. Or you're *Skypeing,* another verbified noun in this new century.

This isn't exactly like using a traditional telephone, but it's pretty close to using a computer and any standard Instant Messaging or email application. You keep track of people's addresses, select them when you want to IM or send them email, and initiate some activity. It just happens, with Skype, you're initiating a phone call.

Calling Skype Users

There are many ways to call Skype users:

- Select the name in the Contacts page and click the green telephone button in the bottom left of the page.
- Double-click the name in the Contact page.
- Right-click the name in the Contact page and choose Call from the context menu.
- Look up calls you made in the past on the Call page and either click the green telephone button on the bottom of the page or right-click and choose Call from the menu.
- Type the name into the text field just above the telephone buttons and click the green button.
- Right-click on the name on your Start page that was left when he tried to call you and you didn't answer.

The first "person" you should call is the Skype Echo Test lady. Her address is echo123 (type it in the text field). Her lovely accent (the consensus among Skype users is that she's Russian) welcomes you to Skype, lets you record yourself for 10 seconds, then plays those words back to you. If things don't work, she offers troubleshooting tips and points you to more help. Lately, other speakers, including computer voices, have performed the echo chores.

Essentially, any place you see a contact name you can initiate a call. Calling is the idea behind Skype, so this makes sense. Figure 6-10 shows the context (right-click) menu on the Contacts page (this is the third method shown in the earlier list of how to call a contact).

You can also see in this figure that there are a variety of other interesting choices for a selected contact.

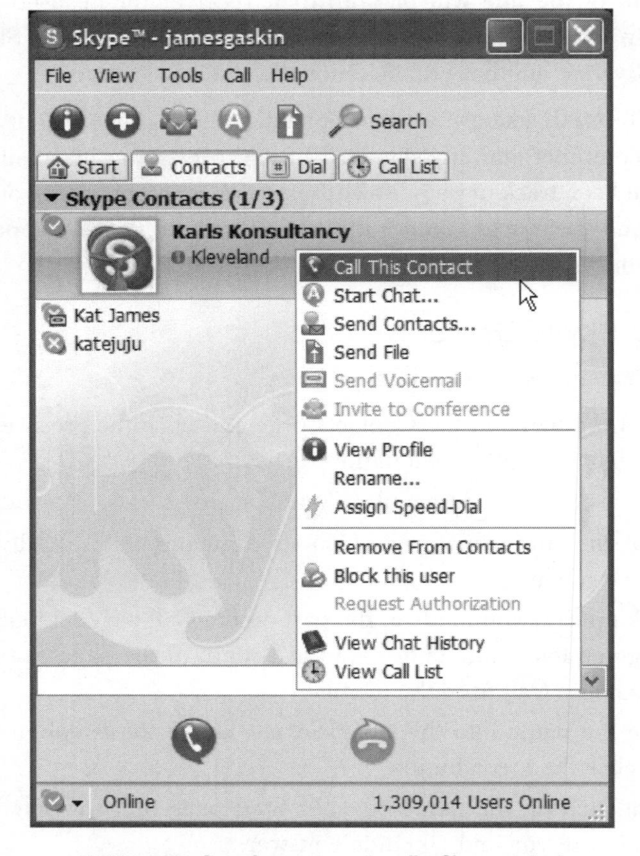

FIGURE 6-10. One of many ways to call a Skype contact

The two contacts below Karls Konsultancy are both offline, which is why their icons are gray (it's hard for you to see the color, but trust me) and marked with an X rather than a check mark. Contact icons also show a clock if they are unavailable, as judged by a lack of keystrokes on their system. If the icon shows a happy face, they are inviting new conversations (called Skype Me). The cassette tape near the top name means voicemail.

Skype's speed in updating the contact status of remote icons amazes me. When the status changes on a contact on your system, the icon changes immediately. The largest delay I've seen is about three seconds (testing from one client on SBC DSL to another client on a completely separate network connected through Comcast cable Internet). That "delay" occurred when another 2+ million users were online. Impressive.

When you call someone, their contact picture takes center stage on your Skype application window. Your name shows up on the callee's Skype, so they know who is calling.

Click the green button with the inset telephone handset to call someone, and click the red button with the handset to hang up. The handset in the red button is horizontal, like it's been hung up already. Even computer-centric systems realize the power of emulating traditional telephone operations.

Chat (Instant Message)

It's silly to call the text-messaging component of a phone application "chat," but that's what Skype does. Blame this on Instant Message applications that started calling their products "chat" to make it seem casual and immediate. It is all that, of course, but chat still means talking in the real world. So rarely do the real world and the computer world line up correctly that we shouldn't be surprised anymore. I say this so you won't click the Chat icon and expect to hear a voice.

If you don't know about Instant Messaging (IM), text messaging, or cell phone text, ask a teenager. Most teens sit at the computer with a dozen or so IM windows running, following a dozen separate conversations while watching TV and doing their homework. At least that's the scene at my house.

Start a chat session just like a call, except choose Chat from the top menu, context menu (the one that appears when you right-click a contact), or Tools menu. Figure 6-11 shows an open Chat window with a session in progress.

Nothing too special, yet, with this chat, is there? But notice the icon at the bottom of the right area. The text beside the blank person icon is "Add more users to this chat!" (including the exclamation point). Multi-user chats can become a nice business, sales, or family update tool. You can also add a contact by clicking the icon on the upper left of the chat window.

Beside the Add icon on the top left is a Topic icon. You can define topics for easy search and archiving. If you add a topic, the name you give the chat appears on a horizontal menu just below the top menu where the Topic icon sits.

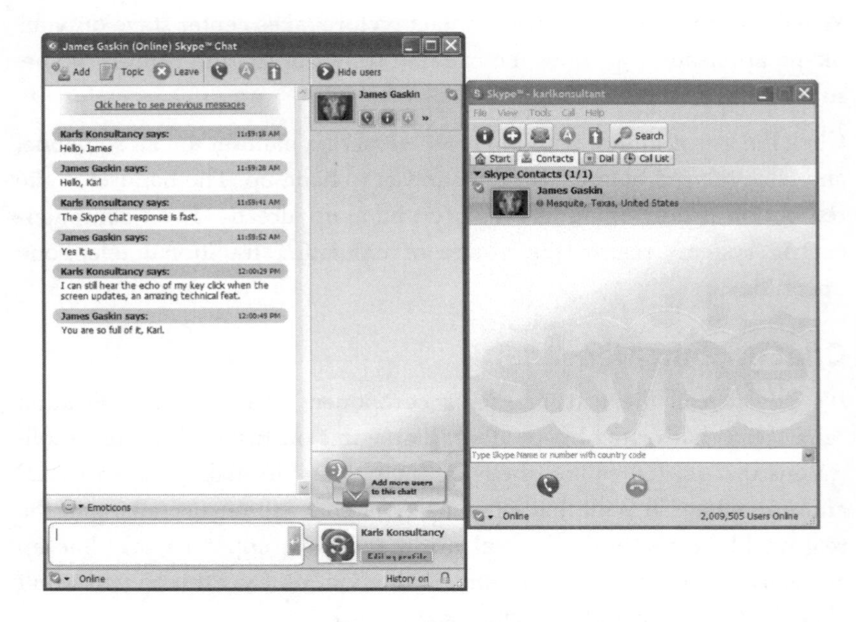

FIGURE 6-11. Skype Chat session in progress

The other icons on the top menu are:

- Leave
- Call
- File Transfer
- Hide users

If you hide the users, the chat portion fills the screen and you can see more text at once. File transfers work slowly, but in the background so you can continue working. You can send a file to one person or click the "Send file to everyone" icon so that all people in the chat receive a copy of sent files.

When you get tired of typing, you can click the Call icon and continue the chat by actually chatting instead of typing. Or you can click the Leave icon and close the session.

When you right-click in the chat area, the context menu offers to Find or Find Again any text you wish. In long chats with multiple people, text search tools save time.

Wish you could remember what you chatted about last week? No problem. One of the options for each client is how long to keep Chat history (File → Options → Privacy tab). You can keep your chat history forever (the default), two weeks, one month, or three months. You can also choose "no history" and trust your memory. This page also offers a Delete history button.

Handy as Chat is, there is no current way to connect to other Instant Messaging services. Skype fans are busy trying to make this work with others, but if you can't wait, check out the built-in Gaim offered by SIPphone.

File Transfer

Files transfer from you to your conversation or chat partner in one drag-and-drop operation, thanks to Skype. It's not fast, and the file won't transfer unless the recipient accepts it manually, but occasionally handy.

You may initiate a file transfer in one of these three ways:

1. Drag-and-drop a file onto a contact icon.
2. Right-click a contact and choose Send File from the context menu.
3. Select a contact, then click the Send File icon (the blue paper with the big white arrow).

If you start the file transfer using the second or third option in the list, a typical file browser window opens so you can choose your file. Click the Open button to start the transfer.

The process isn't fast, and in fact may be painfully slow if the file recipient is behind a router using certain types of security. If the file transfer window says the file is being *relayed*, it will be, shall we say, deliberate. Figure 6-12 shows a file transfer in progress.

While not fast, the file transfer option is handy and works in the background. You can go on talking or using Chat, or just hang up and let the file transfer finish. The manual intervention part happens at the beginning when your recipient accepts the transfer. After that, it putters slowly along without needing attention.

Two good features make this bearable. First, the file, just like a Skype conversation, is encrypted. This may be the easiest way to send files with end-to-end encryption on the Internet. Second, the transfer will complete as long as both partners leave the transfer windows open. Even if one partner gets disconnected or turns off their computer, the transfer will pick up where it left off when both parties are online again.

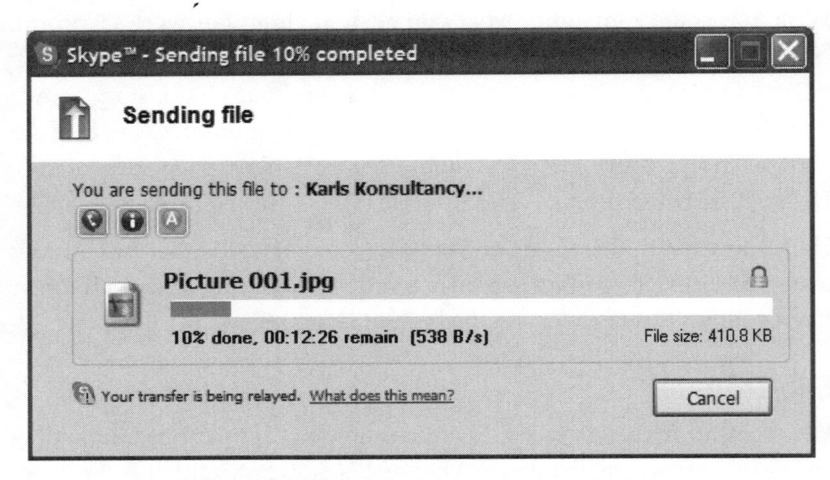

FIGURE 6-12. A Skype relayed file transfer in slow motion

The transfer window, as you can see in Figure 6-12, includes icons to initiate a phone conversation or chat session. Skype nicely warns recipients that files may contain viruses and the like, but you may turn off that warning after the first time.

Client Configuration

Skype provides more configuration tools than most applications, yet most users don't need to change any of them. The best part is almost every configuration option takes away something that may be annoying, so you can make Skype behave as subtly as you wish.

The following images from the Skype-Options pages all show the default settings. When you install Skype, this is how it will be configured. If you make changes and want to undo them later, you can use the following screenshots as a guide.

Figure 6-13 shows the General tab, which handles the startup configuration details such as whether to start Skype with Windows and checking for updates. Since Skype updates require a full download of the new client (around 6 MB) each time they increment even a little, I don't know if I want to check each time I log in, but perhaps I might download it after I'm finished to have it ready for the next session.

I feel the time settings for the Away and Not Available settings are too short. Five minutes comes quickly when you're looking at papers beside the computer, and it can hint to others that you're not at your computer

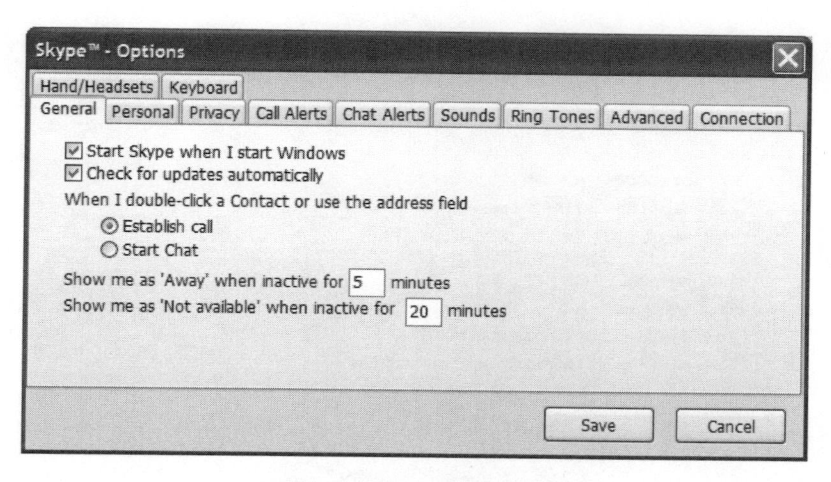

FIGURE 6-13. Skype startup settings

when you probably are. You might try setting the Away option to 10 minutes, and the Not Available option to 12 minutes so people will rightly think you've left your computer. There's not a lot of difference between the two, but Not Available is a stronger warning than Away.

The next tab, Personal, does three things, and only one of them is on this page. There are three command buttons on the Personal page:

Go to my Skype Account Page
: Opens the Skype web page so you can log in.

Edit my Skype Profile
: Provides the same result as choosing File → Your Personal Profile, as shown in Figure 6-8.

Change Password
: This is something you might use. Put in your old password and then type and retype the new password.

Figure 6-14 shows the Privacy page, which allows you to define who can call and/or Chat with you. You can also block Skype members you don't want to ever see again (only in Skype, unfortunately, and not real life).

The "Allow calls from" section default lets anyone call you and offers two restrictive options. If you add a person to your Contacts, they can call you, just like a white list for email to allow known users to bypass the spam filter. Tightening up to the maximum, you can block anyone you haven't specifically authorized. The same options exist for Chat.

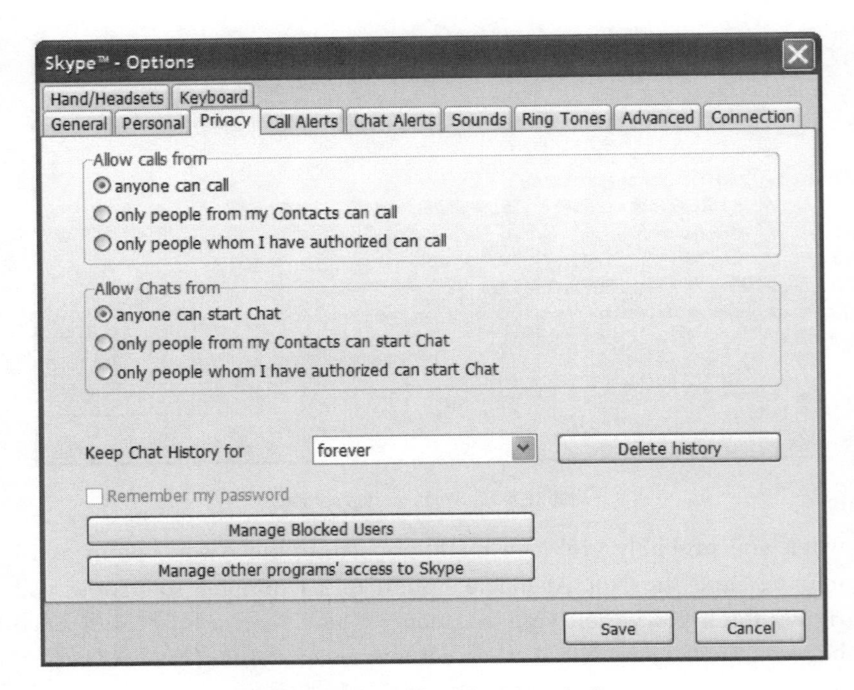

FIGURE 6-14. Controlling who can contact you

Speaking of Chat, the middle section is where you can decide how long to keep a text file of your Chat sessions. Options include "forever" (as you can see), "not at all," "2 weeks," "1 month," or "3 months." Text takes almost no disk space, so unless you Chat constantly, this shouldn't be a worry. And if you Chat about something important, you can easily find the conversation again (highlight the user, then look under Tools → Recent chats).

On the other hand, if you do run low on disk space, you can click the Delete history button and clear off some space, or you can clear the history to hide evidence.

Users can be blocked from the main Skype application. Right-click the contact name. "Block this user" is toward the bottom of the menu. If you change your mind, click the Manage Blocked Users button and release them from non-Contact status (with the "Unblock user" button).

If another program needs access to Skype for communication needs, you can check that status with the last button. Don't worry about this unless a program starts, begins to load Skype, and hangs completely.

I like Skype's options in the Call Alerts tab, because they give you good control over how your system responds when called, depending on who calls you. Figure 6-15 shows this page.

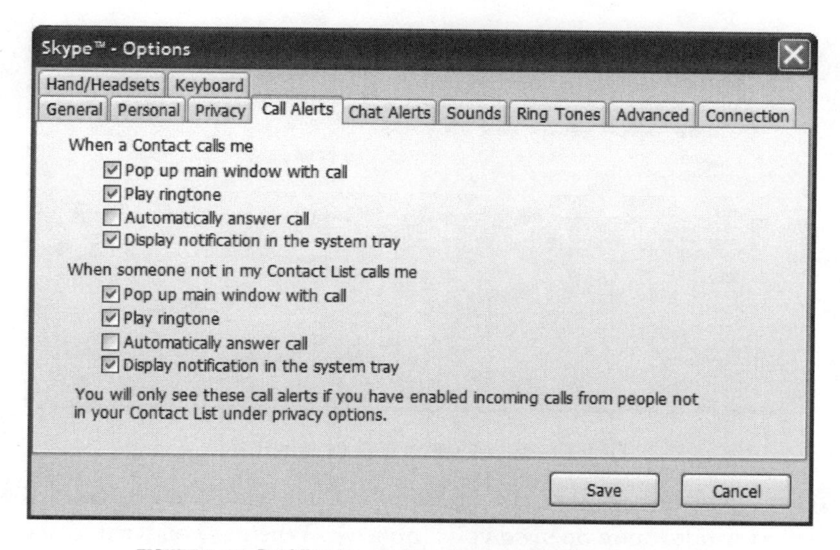

FIGURE 6-15. Deciding how bothered you want to be when called

The defaults treat every caller the same, whether you have them listed as a Contact or not. These options work well if you're using Skype in a business situation and want to react to internal calls (Contacts) differently from unknown callers.

Remember the section in the Privacy tab that lets you block callers under various circumstances. The last line reminds you of the Privacy settings, which is a nice user touch few other applications remember.

One would think the Chat defaults would match those for calls, but they don't. Notice the differences in Figure 6-16.

Where calls aren't automatically answered by default, the Chat window opens automatically, which is the Chat equivalent to answering the call. And Skype likes their Chat so much they use the same warnings for Contacts as non-Contacts.

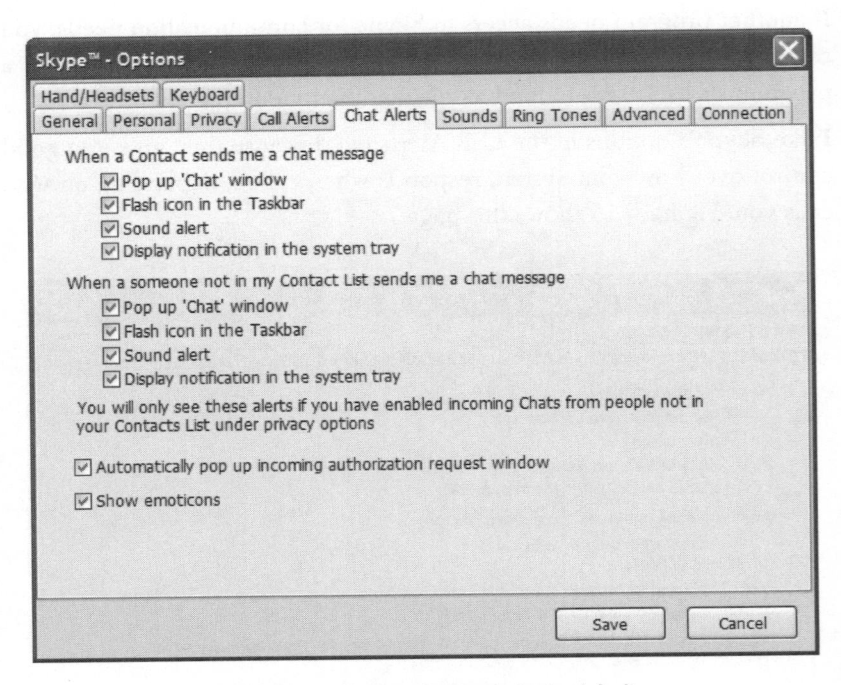

FIGURE 6-16. Chats make more noise by default

When someone sends a file or starts a Chat, you need to have the request window pop open so you'll notice it. Whether you want to show your emoticons (graphical icons such as the smiley face) is a different story (so 1990s).

Skype is an auditory application, and Figure 6-17 shows how noisy it can be.

They probably don't need an entire page for this, but here it is. You might not need a sound when you put someone on hold, because it's an action you take, so you should know it's happening. Knowing when the other people hang up could be handy, however, so you don't keep going, "Hello? Hello?" into your headset and start cursing Skype.

Even noisier, you can specify the ringtones you like. Figure 6-18 shows you can again segregate Contacts from the unworthy.

Any mono WAV file will work in any of these options, although long sounds will mean a large file to load and play. You can find WAV files all over the Internet and inside your operating system, especially if you have a game or three loaded.

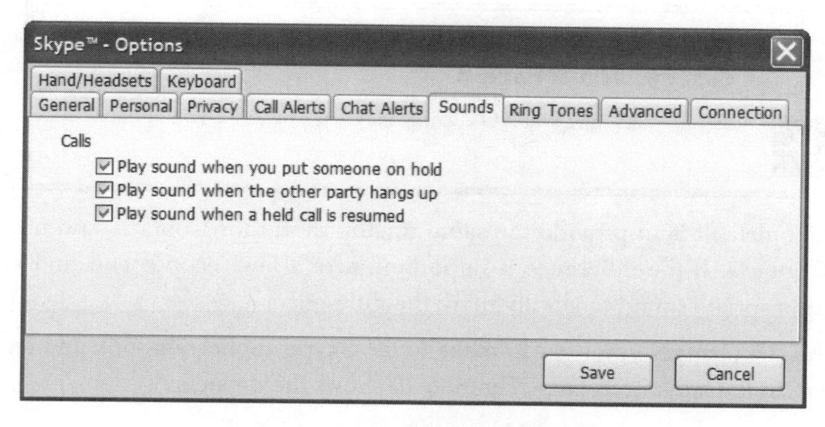

FIGURE 6-17. Making noise with every action

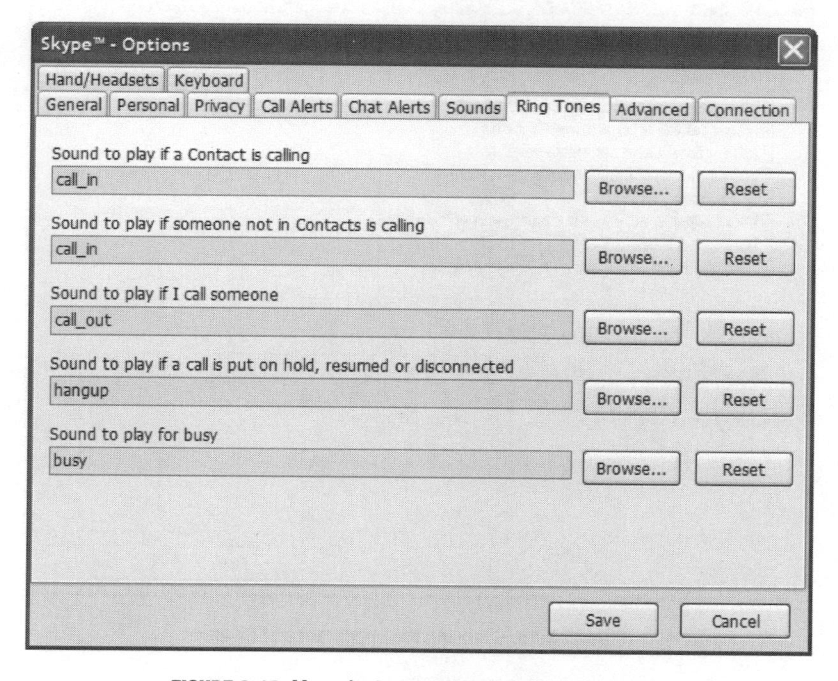

FIGURE 6-18. More ringtone options than your cell phone

Even more fun, you can use the Windows Recorder (and Macintosh makes it even easier) to record your own sounds. After all, the headset or handset microphone will work pretty well for casual applications, especially if they're spoken words.

Sorry, No Stereo

Stereo WAV files will NOT work–they must be one channel only.

The default is to provide the same ringing sound for Contacts and non-Contacts. If the difference is important, take a few seconds and find an appropriate sound to aurally mark the difference.

All the changes you want to make to the Skype application look and feel are in the Advanced page. Figure 6-19 shows the defaults.

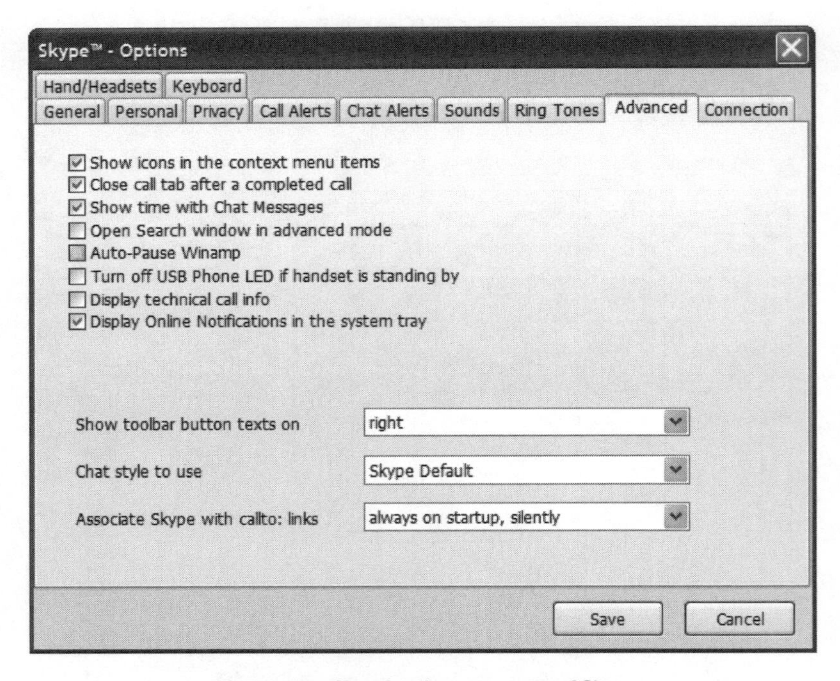

FIGURE 6-19. Choosing the appearance of Skype

Show icons in your context menu? Most applications don't, but Skype does. If that bothers you, uncheck the first box.

I agree with closing the call tab after a call, and showing time in Chat messages. That's particularly handy if you keep and refer to your history files at all.

The default for Search is to open in basic mode, but that never offers enough options for me. I would change the fourth line to "Open Search window in advanced mode".

I really appreciate the fifth option, which is slightly highlighted and is called Auto-Pause Winamp. If you play Internet radio stations through Winamp (I do quite often), Skype will pause Winamp when a call comes in. Some luxury cars automatically mute the radio when the car phone rings, and this offers the same consideration.

The middle drop-down box option on the lower-right of the page lets you change the Chat style. Your other choice is the type of chat listing used by IRC (Internet Relay Chat). Chat programs have been around for a long time, and if you're used to IRC, Skype accommodates you.

Unless you have two telephone applications on your computer, the last option won't matter. If you do have another application, you can tell Skype to defer to the other program by telling it to never respond to a programming command (callto:).

If you have a firewall problem, the Connection page will become too familiar. Figure 6-20 highlights the random port number that is assigned to this computer when Skype is installed.

Skype's technical answers page is *www.skype.com/help/faq/technical.html*. There you will read that all TCP ports above 1024 (outgoing only) should be left open for Skype, and all UDP ports outgoing, and answering incoming packets through those ports, should also be opened. With a business firewall configuration, that's difficult, because businesses want ports blocked (home and small business users will have little trouble unless they have an advanced router). The more outgoing ports open, the easier your Skype will have finding and communicating with other Skype users.

As a second choice, port 443 (in the text box) should be opened along with checking the box to use Port 80 as an alternate as well. Port 80 is always open, since that's the default port for web browsers.

The rest of the page about proxy detection and authentication won't matter for homes and small offices with low-end routers because Skype will handle all this automatically. If your installation is in a business, your network administrator may have to help you with the proxy details. That is, if your business allows Skype, because many businesses don't. Shh.

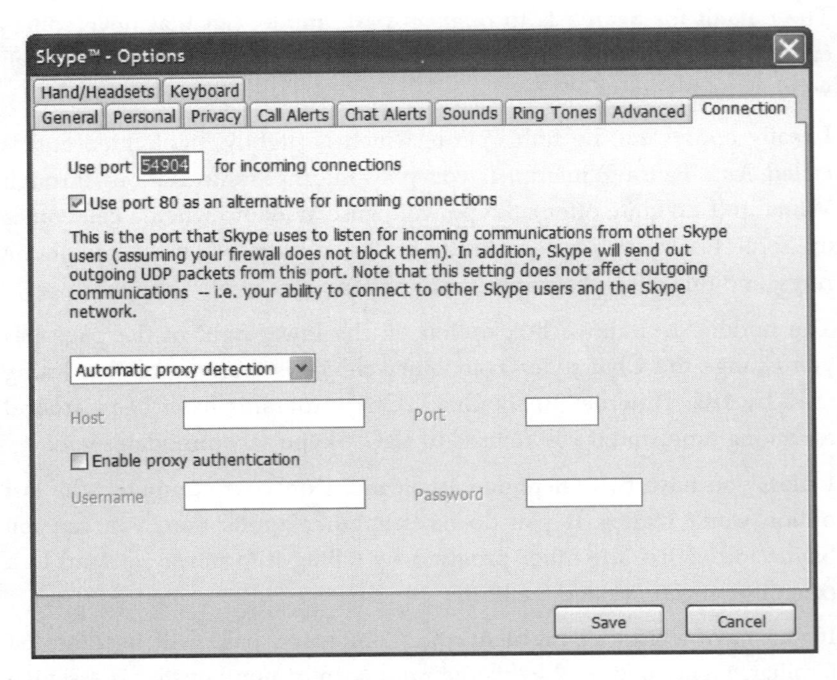

FIGURE 6-20. A page of aggravation if your calls don't get through your firewall

More sound options are shown next, in Figure 6-21, but these sounds cover your ability to hear or be heard.

Each drop-down option will show the name of the actual sound hardware used in your computer. For one of my machines, it's ESS Maestro. For another, it's SiS 7018 Wave. Your computers will have different names, but unless you work with sophisticated audio hardware or programs, the Windows default device will be your best choice.

Every good program offers hotkeys or special key combinations to speed certain functions. A good hotkey can save you a lot of mouse crawling. Figure 6-22 shows the hotkeys that Skype includes.

The default for these hotkeys is off. Since good programs offer hotkeys, the good hotkey choices may already be taken on your computer. If not, you can use their suggestions or create your own hotkeys. Every hotkey used can give your mouse wrist a rest and can be performed faster than using the mouse. And they are more accurate, since you don't have to worry about positioning a cursor exactly on a menu item.

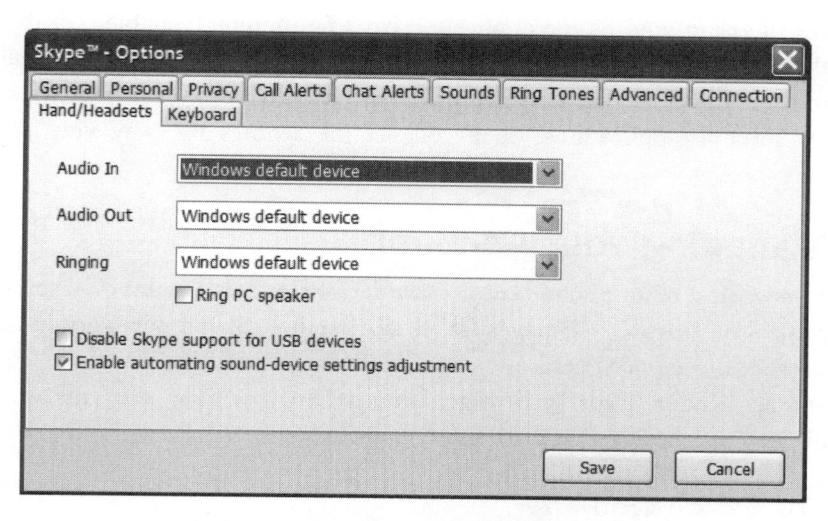

FIGURE 6-21. Choosing the sound conduit for Windows

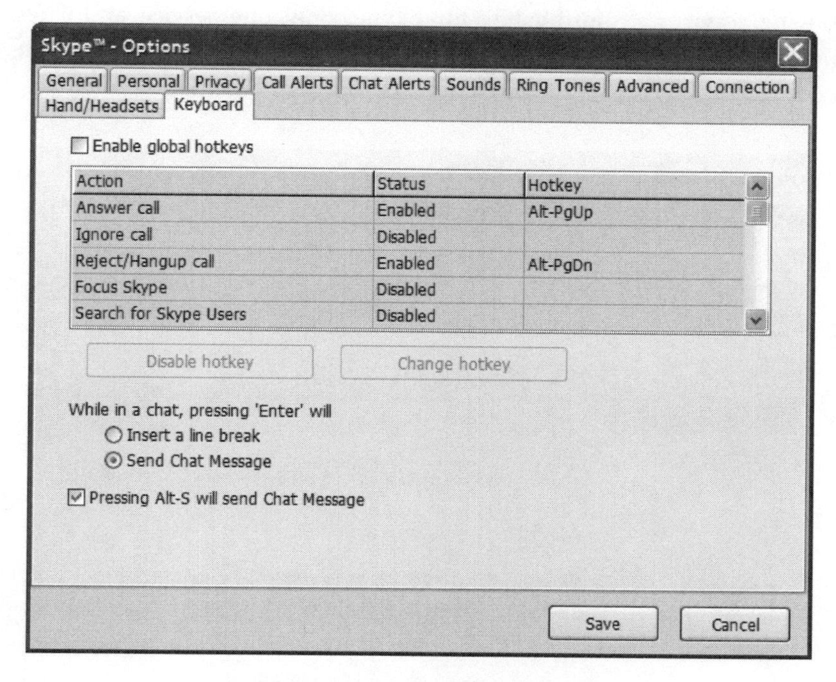

FIGURE 6-22. Key strokes to rest your mouse

Few items in your Skype application can't be modified, enabled, or disabled. The more application control you have, the more you will feel comfortable with that application. Skype should be one of your most comfortable applications, once you get the settings the way you like them best.

Managing Your Account

Vonage and other phone-centric providers include quite a bit of account control on private web pages for each customer. Skype puts very little, because the service details are configured almost completely with the client application. Less web page management by users also fits into Skype's decentralized approach to the business.

The Skype Web Site

Important reasons to go to the web site for your Skype account include buying services. Unfortunately for Skype, only one service and a few products are available for sale right now: SkypeOut and phone headsets and handsets. Figure 6-23 shows the overview screen for an account.

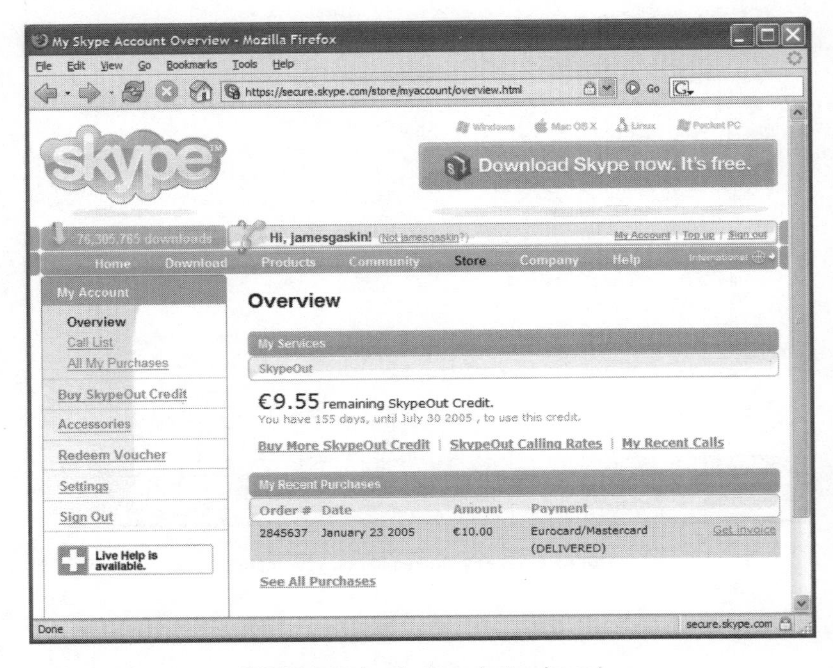

FIGURE 6-23. Log in, buy minutes, log out

You reach this screen by providing your Skype name and password, the same ones you use at the client. Activities you can perform on the web site run down the left side of the page, under the My Account heading.

The SkypeOut page shows two things: all your calls and all your Skype purchases. When a service bills you for calls, they must keep track of those calls so you can argue with them later.

Buying SkypeOut credits works easily and simply unless it goes horribly wrong. More details about that issue await in the next chapter.

Skype generates revenue through SkypeOut minutes and some hardware sales. They partner with a growing list of hardware vendors to include Skype functionality in new devices. Figure 6-24 shows the Accessories page.

FIGURE 6-24. Skype-friendly hardware

Plantronics leads the headset market, and has for years, starting way back before regular phone users ever considered a headset. Their connection with Skype gave the upstart broadband phone company some cachet in the marketplace.

The second two items illustrate Skype's attempt to look more like a phone-centric provider than a computer-centric provider, or at least to give some customers the type of phone interaction that makes them comfortable. The middle phone handset uses a USB connector to link to a computer and the Skype software running on that system. The Olympia cordless DUALphone cuts the six-foot tether that every headset wearer strains against by using a USB connection to link a standard cordless telephone handset to Skype software. Handily, it also works as a regular traditional telephone handset, servicing both types of telephone connections.

Other menu options on the Skype My Account page duplicate client options: change your password, and update your email address. If you have a voucher for SkypeOut minutes, the Redeem Voucher menu item leads you to the page where you type in the voucher number for redemption.

Your Skype Application

While you can change your password and email address on the web site, you can do the same from within your client software. The Personal Profile discussed earlier (refer back to Figure 6-8) shows the email address field to change. To change your password, go to Tools → Options and click the Personal page tab. The third button on that page, Change Password, opens text fields for your old password, your new password, and the new password typed a second time.

That same Personal page includes a button to open your web browser to the My Account page on Skype. Put in your username and password, and you'll see the page shown earlier in Figure 6-23. You can reach the same page by choosing the Tool menu option on your Skype client application and clicking the Go to My Account Page item.

Skype for the Pocket PC

By spring 2005, there were 1.3 million downloads of Skype software for Pocket PC (just over 1% of all downloads). Skype's network administrators can't differentiate between clients on the network, so they don't know how many Pocket PC users have been on the system.

Requirements include Pocket PC Version 1.0 or higher or any PDA using Microsoft Windows Mobile 2003 for Pocket PC. The Pocket PC must have at least a 400MHz processor. A headset is strongly recommended. You will still annoy people around you when talking, but at least you won't have to yell at the face of your PDA hoping the microphone picks up your voice.

The Pocket PC Phone Edition includes a data service (provided by the cell phone company) called GPRS (General Packet Radio Service) or 1xRTT (Single Carrier Radio Transmission Technology, as used by Sprint and Verizon). These are fast enough to support Chat (IM) but not voice calls. You must have Wi-Fi support in order to have enough bandwidth to support telephone calls. (But if you obtain a Pocket PC Phone Edition with 1xEV-DO or EDGE, both part of the latest generation of cellular data, you may find it's fast enough to support Skype–but, check your service plan first to make sure you're not being billed by the kilobyte!)

You must be a registered Skype user before installing the Pocket PC software. Skype software installation for a Pocket PC requires three steps:

1. Download and install the PC portion of the software (*SkypeForPocketPC.exe*).
2. Verify ActiveSync is running on your PC.
3. Download and install the Pocket PC software (*SkypeForPocketPC.cab*) to your handheld device.

Once installed, your Pocket PC becomes a Skype client just like any other (for the most part). If you log into both a desktop and Pocket PC concurrently, both will ring when you get a call. Answer from your Pocket PC and carry on just like any other conversation. When you call out from your Pocket PC, you can use SkypeOut minutes.

There are a few caveats:

- You must recreate your contact list because no copy or transfer option exists.
- You can participate in conference calls, but you can not initiate them.
- Only English language support is available (as of Spring 2005).

Be aware the level of standardization taken for granted with PCs running Windows software does not yet extend to the Pocket PC world. Skype's Pocket PC forums are full of unlucky customers searching for the right Pocket PC operating system, Wi-Fi support, and Skype software version to get full functionality.

Bluetooth headsets introduce another level of complexity into the equation. While handy, and sometimes the only way to get a microphone input for some Pocket PC models, finding the right combination of headset, Pocket PC, and Skype version for that Pocket PC can be tough.

Manage your expectations with Skype for Pocket PC. The list of models with successful Skype support isn't long but is growing. Unfortunately, Skype does not list "approved" models but instead points customers back to the requirements list.

If you're planning to buy a Pocket PC for Skype use, go to the Skype Pocket PC forum (*forum.skype.com/viewforum.php?f=13*) prior to purchase. The same goes when you're shopping for a Bluetooth headset. Hearing a success story or two before buying will smooth the process because you know at least a few other people achieved their Skype and Pocket PC goal.

What Skype Forgets to Tell You

Skype emphasizes the "easy" tag for Internet Telephony, and they do a great job. But there are always things they forget to mention. Unlike Macintosh fanatics who don't dare speak ill of the Mac, Skype users will speak up about the shortcomings they fight.

Technical Details They Don't Mention

The following list describes the technical elements Skype doesn't specify:

1. Standards exist, but Skype doesn't subscribe to them. They only mention the world of SIP phones deep inside a few of the Frequently Asked Question web pages. While the Internet oversight committees work on the technical details to advance Internet Telephony, Skype goes their own way.

   There are two ways to set standards in the computer world: agreement among all involved and market share. Skype jumped out so fast and gained so much market share that they are setting the standard. But coordinating the technical details needed to call from one computer-centric phone network to another demands agreement from all involved vendors.

   SIP phone networks are working and have demonstrated success in calling between competitors. After all, if you and a friend are both using softphones on the Internet and don't touch the traditional telephone network anywhere, connection should be simple. It's not yet, and unless Skype changes, it may never be simple.

Unless, of course, the whole world switches to the Skype model. Not likely, but the idea that one operating system, Microsoft Windows, would control 90% of the personal computers in the world seemed silly at one time. Now Microsoft controls the PC world, and Skype's user base has an even greater percentage of the total computer-centric broadband phone market.

2. SkypeOut quality can best be described as spotty. Random factors cause lousy connections too often, and some countries drop off the SkypeOut map entirely at times (anyone call France successfully in January 2005, especially with Pocket PCs?). Telephone network integration takes hard work, but other broadband phone companies provide excellent call quality and coverage, so SkypeOut must keep up.

3. Security issues are appearing and will get worse. Skype has a terrible reputation with large companies who find it unacceptable to let the service through their firewalls so easily. What is a clever technology solution for Skype–the relayed file transfers and other way through firewalls–causes heartburn and sleepless nights for security administrators.

   The file-transfer ability included with Skype magnifies the problem. Files that are transferred through the Skype holes in a company's security net bypass every virus check. In other words, any Skype file transfer includes the potential to completely infect the company network, with no record of where the infection entered the company.

Business Details They Don't Mention

The following list describes the business elements Skype doesn't specify:

1. Money, and the need for money to support Skype, will become critical over 2005. They have millions of venture capital investments, including $19 million in March 2004 as a second round. Executives refused to say how much their initial funding was, but said it was "less than $10 million" (Zennstrom interview with CNET 12/2/2003).

   Admittedly, the peer-to-peer model reduces costs, and the viral marketing scheme worked so well I don't think they've spent any significant advertising dollars. But the services they must provide to keep growing, such as SkypeOut improvements, SkypeIn, voicemail, and their hinted Skype for Business, will require investment.

Right now, the number of Skype registered users paying for SkypeOut hovers around three percent. Of 100 users, 3 pay something. That's not a recipe for financial success, even with a million paying customers (when they're supporting many millions of nonpaying customers).

2. SkypeOut payments must improve. The Skype forums burst with complaints about the third party authorized early on to handle credit card transactions, MoneyBookers (*www.moneybookers.com*). Plans to accept PayPal transactions are underway but not enabled or available in early 2005.
3. Converting customers from free services to paid services takes careful management and strong execution. Many products make that transition, but many more fail and disappear. Skype's strong market position will keep them afloat if they make mistakes because of customer loyalty. But every other broadband phone company now targets every person with a broadband Internet connection, giving customers alternatives to Skype if they drop the ball.
4. SkypeIn must work perfectly if Skype will ever be used as a primary phone and keep growing. How much would you pay to let a Ma Bell customer call you? You pay Vonage $15 or $25 per month for that privilege. Will you feel comfortable paying Skype 5 euros every three months? Will that convince you to switch to a computer-centric phone service for your second line for long distance cost reduction? Skype bets their company you will, and that's a big bet.

Troubleshooting

Most help requests revolve around sound quality. Cheap headsets are better than computer speakers with a microphone, but they aren't great. If you can't get a headset, see if an old pair of headphones from a portable CD player can replace your computer speakers during calls, because that will help considerably. But good headsets with a built-in microphone offer the best sound quality.

There may be a reason SkypeOut support appears at the top of the Skype "Getting Help for Skype" page on their web site. When you can't get satisfaction, especially when dealing with Moneybookers, join many others with similar tales of woe in the Skype Forums. It won't solve your problem, but at least you won't feel alone.

Skype does an excellent job using animation on their web site to help new users configure Skype and their computers to support Skype. Go to *www.skype.com/help/guides/* to start one of the dozen animations and see for yourself how to handle configuration issues.

The primary Skype support forum (*forum.skype.com/viewforum.php?f=2*) includes thousands of user questions. I can't decide if it's good that so many people go to the source to find solutions, or if it's bad because so many people have trouble. But Skype claims over 20 million registered users, so a few thousand messages in the help forum constitutes a tiny user percentage.

Windows computers support the vast majority of Skype clients. Windows can be a rather, um, interesting support issue itself (see many Windows support and troubleshooting books at *www.oreilly.com*).

Here are my recommendations when you have Skype problems:

- If Windows has upgraded, even a little, upgrade and reload Skype. Upgrades will retain your contact lists and other configuration details.
- Hang up and try again. Internet traffic, and therefore quality of voice connections, varies depending on many factors. Add a connection to a traditional telephone line, such as when using SkypeOut, and the potential disaster points multiply. So just shrug and try again.
- Upgrade the sound drivers on your computer, especially when you have installation problems getting sounds in and out of Skype.
- Persistent sound issues may be investigated on the Skype Sound Set Up Guide web pages (*www.skype.com/help/guides/soundsetup.html*). They provide excellent step-by-step instructions on configuring Windows, Linux, and Macintosh systems along with your Skype client.
- Reboot your computer. Rebooting solves many bizarre problems that nothing else will fix. If rebooting by restarting doesn't work, turn off the computer and all connected peripherals for one minute and try again. Remember to unplug USB devices if things stay screwy, because they can sometimes remain hung up if still connected to the computer.
- Verify your broadband connection works as expected. You can't blame Skype if your service provider can't find the Internet.

- Reboot your broadband modem and router. Turn off everything, wait one full minute, then:
  a. Turn on your cable or DSL modem and wait until it connects to your service provider.
  b. Turn on your router and wait until it connects to your router properly.
  c. Turn on your computer and wait until it connects to your router properly.
  d. Start a web browser and verify you can surf the Internet properly.
  e. Turn on Skype and try again.

Redial

Millions love Skype, and for good reason. A free phone call anywhere in the world to anyone with an Internet connection is the type of disruptive technology that scares incumbents (AT&T, the monolith, disappears when purchased by SBC) and thrills users (100,000 or so new Skype users per day, thanks to word of mouth).

But computer-centric broadband phone services, whether Skype or one of the SIP phone companies, will satisfy only a small percentage of total telephone users. The telephone in its traditional form and function resonates strongly with most people.

If you have friends far away with Internet connections, Skype puts you together for the least money possible (free is good). But if you feel more comfortable with a phone-centric service as discussed in the previous chapter, you have nothing to apologize for or feel guilty about.

The best part? You can mix an inexpensive phone-centric provider to get away from the traditional telephone company and use Skype to call friends outside your free calling area. Or, you can use Skype from your laptop when you travel and call back home.

Monopolies on telephone providers and services disappeared. Use Skype, use SIPphone, use FreeWorld Dialup, use Vonage, use Packet8, use BroadVoice, use your cell phone. Use whatever you want for the calls you want, and enjoy.

7
911, ALARMS, AND OTHER OUTGOING CALLS

Outgoing calls from phone-centric providers (Vonage, et al) come with the service package, because they emulate traditional telephones. People expect any traditional telephone they pick up to reach any number they wish to dial, which the phones do, even if the first leg of the trip runs over a broadband cable instead of telephone wires.

However, behind the scenes, the telephone network evolved and continues to function in some bizarre ways. Technical systems built by traditions rather than modern engineering principles include a variety of quirky features and downright weird processes, and the traditional telephone network carries their bizarre process banner high.

So even though a phone-centric provider lets you pick up a phone and dial Aunt Martha, reaching a 911 operator starts an entirely different (weird) process. Those of you with monitored alarm systems may also need to take a couple of extra steps.

911 calls will be an issue for phone-centric providers for the next couple of years, but this wouldn't be a big problem if the broadband phone companies had planned better. Government committees are looking into the problem now, which will likely muddy rather than clarify the issue.

Computer-centric phones involve an entirely different set of problems and will lag far behind for years and years. Those of you with softphones who want to call 911 from your PDA may not reach that goal for another 10 years. OK, maybe five years.

911 Issues

Taking the long-term view, 911 calls over broadband phones should be sorted out by 2007 or 2008. Growing pains will cause some awkwardness and make too many tragic headlines as broadband phone companies struggle to improve their technology and gain access to emergency service centers now controlled by traditional telephone companies. Internet Telephony providers, or at least the phone-centric group, will be forced to start paying into the pool that supports 911 service, often labeled "Cost Recovery." Currently, traditional telephone line users are charged under a dollar per month per line (many are actually a half dollar or less, but that will likely increase).

The 411 on 911

Curious? A real 911 fan? Check out the National Emergency Number Association at *http://nena.org.*

Some seem to think 911 arrived long ago, completely functional, and one call automatically delivered many vital details to an emergency operator from the beginning. Nope. The initial push for what became 911 was in 1957, when a national group of Fire Chiefs recommended a single number to report fires. Thirty years later, only half the phones in the United States had any type of 911 service. Even in 2000, about 10% of the U.S. population who called 911 still had to provide their location, because the emergency operator did not receive that information from the system.

The automatic delivery of a caller's phone number and physical address to the 911 personnel receiving the call is called Enhanced 911. You may see this abbreviated as E911, e9-1-1, or e911.

All the noise about cell phones and 911 calls? Even though cell phone providers are "real" telephone companies in every sense of the word, only about 40% of the emergency operators in the U.S. can receive E911 calls from cell phone users. A cell phone user can call 911 just about everywhere in the U.S. and reach an emergency operator, but only with what's called Level 1 access.

This means no phone number or address information will be delivered to the emergency operators. Broadband phone services, especially the phone-centric vendors, should at least have this level of support, but they don't. The big push to deliver Enhanced 911 calls via cell phone has yet to provide such support for even half the U.S., but the cell phone industry faces different technical problems because their clients are always mobile. Broadband phone services have very few mobile customers and can't hide behind that excuse.

Broadband phone providers, by the decentralized nature of their service, struggle with traditional 911 service connections.The following list provides some reasons why broadband phones have more trouble than traditional telephone providers delivering 911 service.

- No phone number is tied to a physical telephone line termination, so providers don't automatically know where you're located.
- You can move your broadband phone router.
- Computer-centric solutions can be anywhere through wireless connections (no easy answers here).
- Your broadband phone provider's switching equipment to route your call may be thousands of miles away from your physical location.

If after more than 40 years after 911 rollout, 10% of the traditional telephone line users are still without Enhanced 911, what hope do the broadband phone providers have? Actually, quite a bit. The technology behind broadband phones, and the improving technology in emergency centers handling 911 calls, will come together in some interesting ways in the next two years.

Plague on Both Houses

On one hand, we have the traditional telephone companies controlling the 911 technology and often blocking broadband phone company access, and then fueling the hype when tragedies occur because of broadband phone service 911 failures. Shame on them. On the other hand, we have broadband phone companies who have not done enough to provide even minimal emergency response support. Heap blame on both sides, because they both deserve plenty.

Broadband Phone Providers and 911 Today

No phone-centric broadband phone providers have the same level of 911 support as traditional telephone companies. Some pilot projects are underway, and some other projects will get started, while the various telephone players coordinate properly.

Back in Chapter 5, I listed 911 support as one of the features in a comparison table. Here is a bit more about each of those companies, and a couple of others not mentioned earlier:

Packet8

They lead the market by developing Enhanced 911 services that route customer calls to the proper emergency operator for their area and send along phone numbers and addresses.

Vonage

They support 911 calling for all customers, even though Enhanced 911 is not available. One of the market leaders with a pilot project in Rhode Island that does support Enhanced 911 for customers who have activated their 911 calling feature.

OptimumVoice

They offer complete Enhanced 911 support, because they come to your home and physically tie your broadband router and telephone adapters to your location. You can *not* take your router to another location as you can with Vonage and other services.

Lingo

They support 911 calling for all callers, but only to connect you to the general emergency line where you must provide your information.

ATT CallVantage

They support 911 for all customers, but again just connecting you to the general emergency number where you must provide your information.

MyPhoneCompany

They don't offer 911 service at this time (they say to keep your traditional telephone line).

BroadVoice

Nothing yet, but they are working on their own "BV911" service, which requires you to register, and then provide an emergency operator with your phone number and location.

VoicePulse

They believe a 911 service on a broadband phone should work exactly like one on a traditional telephone line, and since it doesn't yet, they don't offer any. Guess they should talk to Packet8 and Vonage.

Emergency operators work in a PSAP (Public Safety Answering Point or Public Service Access Point). These centers can be run by the local police or fire department, ambulance service, or a facility run by a coordinating body, such as a city or county. 911 calls come into the center, connect to a database of numbers and addresses called the DMS/ALI (Data Management System/Automatic Location Identification) system, and provide that information to the emergency operator.

Character Flaws

PSAP with DMS and ALI? Telephone people may be even crazier about acronyms than computer people.

These emergency centers also accept calls through lines other than just their 911 connections. This is how most broadband phone companies connect, because they are not always authorized as official telecommunications providers with the right to connect into the 911 system. Because of the seriousness required for emergency responses, there are tight restrictions on system access.

Because computers run 911 centers, and computers run broadband phone companies, coordinating the two requires more meeting time than installation time. Michael Tribolet, Executive Vice President of Operations at Vonage, said the Rhode Island trial took months to set up politically but only 30 minutes to install technically. Vonage worked with Intrado Inc. (*www.intrado.com*) in Rhode Island, and Intrado provides 911 services to many communities. Tribolet made it clear that Vonage wants "real" 911 service, so they called one of the leaders in 911 technology to help them.

Packet8 worked with their primary network provider, Level 3 Communications (*www.level3.com*), leveraging their existing data centers all over the U.S. By grabbing their customer's 911 calls as they're made, adding the phone number and location information necessary, Packet8 can route those calls through the 911 inbound connection. Emergency operators see the same information from a Packet8 customer as they would from a traditional telephone line user. All Packet8 service plans receive this Enhanced 911 support, but some geographical areas may not yet be activated.

As with every developing market, some lead and the rest follow. Packet8 and Vonage are leading, and the others are following, but everyone will integrate "real" 911 service over the next two years.

How Broadband Phone Providers Should Be Handling 911 Calls Today

It has been irresponsible for broadband phone companies to pitch phone service to consumers and drop the ball completely on 911 support (like the three companies mentioned earlier who provide no 911 service options). Technical issues aside, the first piece of paper a consumer sees when opening their phone-equipped router or telephone adapter should say in big letters "911 CALL SETUP DETAILS." That same headline should be at the top of the web configuration page you use to set up your service. Cost for these changes? Pennies per customer.

It's almost as if broadband phone companies don't really expect people to replace their traditional telephone line with the new service. And the advice to call 911 on your traditional telephone stinks, too. Every phone must support 911.

When broadband phone providers ship a router or telephone adapter, they know your address. They just need to track that address in a database. When you buy a router at retail, you must provide an address to begin your service. Broadband phone providers should institute a very simple rule: if no physical address, then no broadband phone service.

Here's what the broadband phone companies should do:

1. Track the physical address where they send the broadband router or telephone adapter, or get it from the customer when they configure the system. Yes, one broadband phone feature allows customers to take their routers with them and call from anywhere they can connect to a broadband network. Realistically, very few people do this, and you can filter for these rare occasions with a little planning.
2. Grab every 911 phone call as it's made and route it to a special operator. The larger providers like Vonage can have their own operators. Smaller companies can pool together and have a central operator station that services all their combined customers. Broadband phone data transport speeds allow you to grab a phone call starting in Miami, route it to Seattle for processing, and complete the call to Cleveland with no noticeable delay.
3. When the broadband phone operator answers a call, they can verify through network addressing that the call originates on the broadband network servicing the physical address the router was shipped to. For backup, the operator can ask the caller to verify their physical address.

4. Finally, the broadband operator should connect the call to the appropriate local response center. Sending the address data directly to the 911 center's computer works best, but verbally relaying the information adds only a few seconds to the call. If traditional telephone companies continue to restrict access to 911 center computers for data transfer, alert the media and put the blame where it belongs.

Technically, these plans take only days to implement. They will cost money, but far less money than fighting one of the lawsuits already filed over 911 problems. Personally, I'd much rather see money spent on improved technology than for lawyers, because improved technology always costs less and provides more benefits than lawsuits.

911 Everywhere?

On May 19th, 2005, the Federal Communications Commission issued an order "that certain providers of voice over Internet protocol (VoIP) phone service supply enhanced 911 (E911) emergency calling capabilities to their customers as a mandatory feature of the service." Moreover, the FCC demanded that traditional telephone companies allow broadband phone service providers access to their 911 systems and decreed this be done within 120 days. Details are vague, so expect active lawyering on all sides. And Skype already issued a statement saying this doesn't apply to them, so even deciding what is and isn't a "voice over Internet Protocol phone service" will probably wind up in court too. But at least the hammer dropped and the integration begins.

Broadband Phone Providers and 911 Tomorrow

Tomorrow is another day, of course, but this tomorrow is several years in the future. I'm guessing four, but that could vary by one less or three more depending on the costs and resistance by traditional telephone companies.

Here's the great advantage of Internet Telephony as it applies to 911 service: once a phone call becomes nothing but data, much more data can be attached, referenced, or indexed by that data. Computers talk to computers faster and more accurately than any two operators ever could.

Many 911 centers gather extra information about homes and businesses in their area as part of emergency preparedness. Fire departments want to know if a company uses hazardous materials, and a building floor plan helps rescue efforts. Gathering, tracking, and presenting this information to rescue workers varies widely between jurisdictions, however, and there are no standards in place.

But as we've seen, standards groups love Internet Telephony. They also love data exchange formats so companies can transmit and receive data in a usable structure, even if they don't really know what's coming. Add these two trends together and interesting things appear.

With more integration and customer preplanning, 911 calls over broadband phones in the future may include the following:

- Complete personnel list of the family or employees
- Floor plans
- Automatic notification of other family members through Instant Messaging or prerecorded voice messages from residents
- Medical information of all family members or employees
- Messages to the property owner or lien holder
- Messages to the insurance provider of record

Any information about your residence, family, or business location that's accessible via computer can become part of the message stream. There's a limit to how much information an ambulance needs, of course, but sending other information to related parties will save time and smooth recovery from the emergency event.

Calling 911 from cell phones remains a bit dicey, but is getting better. The broadband phone equivalent, using a softphone from a laptop or PDA, won't get better for quite a while, although the convergence of wireless networks and cell phone technology may speed things along. GPS support in all mobile devices would really help.

There's no easy way to track location because the wireless data network doesn't work the same way the cell phone network does. A cell phone provider handles all their customer's calls for an area, so they know if the cell phone is near because they keep tabs on it. Wi-Fi networks come from a wide variety of different vendors, such as coffee shops, paid access areas like airports, and poorly managed private networks open to any wandering user searching for a signal. No central coordination of wireless data access means no way to gather client information and cross-reference the softphone number with the wireless client's identity.

So if you plan on creating an emergency for yourself, just remember that a softphone makes it hard to call for help. Don't throw out that cell phone just yet.

Alarm Systems and TV Device Phone Links

Certain modern appliances like to "phone home" for a variety of reasons. Most common are monitored alarm systems that use a telephone link to report a problem. Satellite TV and digital recording services such as TiVo, that allow you to order Pay Per View movies by clicking your remote also have a phone connection to send the request upstream and receive the authorization back from the TV provider.

If you forget about these services and replace your existing traditional telephone service with a broadband phone, you may have an unpleasant surprise one day. I just hope it's the lack of a movie rather than a lack of police response if your home is invaded.

Alarm System Advice

The problem hits when you realize your monitored alarm system panel–perhaps in your master bedroom closet–ties into your home's phone wiring somewhere, but you have no idea where. During the alarm installation, the tech or techs found the house phone wires and tapped into the line somewhere. The tap may be near the control panel, it may be near the phone junction box where the phone lines enter, or it may be in the attic where they strung the wire for your alarm sensors.

You will rarely find the point where the alarm system connects, and you will probably have to pay a service charge if you ask a technician to come and find it for you. But it may be worth it, depending on the type of alarm you have and your reliance on monitoring.

If your alarm phone line can be rerouted to connect to your broadband router, you're in good shape. That's a big if, of course. If you get lucky, you can go to Radio Shack and buy a splitter that connects two phones to one plug and so link your alarm system as well as your telephone to your broadband router.

There's a caveat with this. To send alerts, your alarm system must use a method that is compatible with broadband networks or voice over broadband. Talk with your alarm company and see if they have any answers for you. Ask if their equipment has been tested with broadband phones. Even some traditional modems used as dialers back to the service have been configured to handle a broadband connection.

Drilling down, there are various types of command-calling protocols, some of which don't work across broadband connections. Again, check with your alarm company.

You have two options, assuming your alarm company understands broadband phones at all. Two of the major industry players, ADT and Brinks, both handle this issue. Be aware they like to call it "VoIP," so use that term if they don't catch on when you say you have a broadband phone.

ADT offers a customized cell backup that connects to the monitoring center through radio frequencies. This customized cell phone provides all the necessary connections for 24-hour monitoring in locations without a traditional telephone line.

Since I'm an ADT customer, I called and asked how much this service costs. The unit and installation cost $249, or $269.54 with sales tax in my area. My monthly bill will increase by $8 if I order this cell backup option.

Surprising Burglars

One security expert offered this interesting thought: if a smart burglar cut your phone line to stop the alarm-monitoring system, but you converted the system to run over your broadband connection, your security system would still be active.

Brinks will install the alarm-monitoring phone line directly to my broadband router, no matter where it is, with a new system. Existing systems will require a service call to retrofit, and that price varies. They're confident their system works over a broadband line and doesn't need to dial a modem.

They also have a cell backup unit, but as a backup only, not just to replace a traditional telephone line. Since I'm not a Brinks customer, they wouldn't tell me the exact price, but it's in the hundreds. I bet if I wanted a cell backup for my mansion when smart burglars cut the phone lines, they would set that up for me. Now I just need a mansion.

Some other interesting factoids about alarm systems and broadband:

- Some alarm systems use pulse dialing (like rotary phones). Those won't work on a broadband phone line. Check your manual and switch the dialing method to tone dialing.
- Alarms use a "seize the line" command to grab a line that's busy when necessary. Broadband phone lines may not respond to that command.
- Be prepared for smaller alarm companies to try and warn you off a broadband phone, most likely because they can't handle the change.
- If you lose your power without a battery backup on your broadband router and connected equipment, you will lose your monitoring. A traditional telephone line receives power from the telephone company, and your alarm has a small battery backup to handle such situations.

As always, when something new comes and upsets established companies, new companies jump in to take advantage. A company called Next Alarm (*www.nextalarm.com*) not only handles broadband phone connections for alarm monitoring, they also embrace Internet connections for monitoring no matter what. They will happily take over monitoring your system via the Internet if you drop your traditional telephone line.

On the other hand, leaving a traditional telephone line in place for around $20 per month (your mileage may vary, of course) solves, or at least sidesteps, the alarm issue. It also sidesteps the Satellite TV and TiVo issue discussed in the next section.

Satellite and TiVo Phone Link Advice

If you have a TV-attached device that offers Pay Per View or other interactive features through the use of an onscreen menu and your remote control, that device uses a phone link or a modem. You probably forgot that the installer ran a phone wire to a plug, because the phone connection is behind the entertainment center and you can't see it.

Dish Network and DirecTV, the two major satellite TV providers, use the phone links to order Pay Per View events. Both offer alternatives, but each alternative includes a premium over the event costs:

Dish Network

$1 for phone and Internet orders

DirecTV

$1.50 for using automated phone ordering, and $5 to talk to someone on their toll-free number

If you do a lot of Pay Per View ordering, these extras could add up. Whether they add up enough to make it worth keeping your traditional telephone line is up to you, not me. If you have an alarm and some Pay Per View devices, the workaround options start to add up to the same cost as a basic traditional telephone line. Just start thinking of that line as your low-tech data link rather than a voice line.

TiVo, the well-known Digital Video Recorder, addresses the broadband phone connection issue directly: "Some customers have reported success with voice over IP systems, however we do not currently support this." Hmm.

However, they continue on and admit that with a Series2 standalone TiVo Digital Video Recorder, you can use a broadband connection after you have completed the Guided Setup process. So the trick is to get your TiVo set up properly before switching over to broadband phones. If you can't manage that, borrow your neighbor's phone with a really long extension cord to reach their phone plug.

Forums for the TiVo community (such as *www.tivo.com/4.3.asp*, *www.tivocommunity.com*, and *www.tivotechies.com*, among many) offer help on this issue. In fact, they offer help on an amazing number of topics, so don't start searching if you're in a hurry. I guarantee some messages will pull you in and steal an hour before you know what happened.

SkypeOut

July 24, 2004, was the day the Skype users of the world could connect to the non-Skype users. SkypeOut, their name for calling traditional telephone numbers, went live that day. Before SkypeOut's first birthday, the service had over a million paying customers.

This step required Skype to get involved with traditional telephone suppliers. They leveraged the modern expertise of Level 3, Cable and Wireless, and other global telecom suppliers who all connect to phone networks all over the world.

Signing Up

There are no extra hoops to jump through to add SkypeOut service to your existing Skype account. The only requirement is to purchase SkypeOut credit, because this works like a phone card. You buy some traditional telephone line connect time via SkypeOut, and then you can call your non-Skype family and friends.

Figure 7-1 shows the Skype web page used to purchase SkypeOut credits. When things work, they work well. When they don't, well, that's discussed next.

FIGURE 7-1. Buying SkypeOut credits for the first time

This page starts the buying process. The default purchase amount is 10 euros, which is about $13.25 depending on the exchange rate. Skype started in Europe and has no American offices, so euros are their currency of choice.

They will ask for various information in order to verify your credit card account. But they quit taking credit cards directly in February 2005 and now only accept MoneyBookers (*www.moneybookers.com*) and PayPal.

Adding PayPal, although only a beta test in February 2005, makes a great deal of sense and answers one of the most serious and consistent complaints about Skype: bad transactions and worse customer service through MoneyBookers. The ranting and raving about MoneyBookers on Skype's own customer forum rolls on forever and ever.

There are few, if any, happy endings with MoneyBookers. People recount being refused transactions on paid-up credit cards regularly accepted by local and online merchants. People recount finally getting the Skype credits they purchased, but it took so much time and aggravation that they still seethe. People recount having purchases disappear, never to be heard from again, or used for SkypeOut credit.

PayPal's chance to reclaim some Skype goodwill should be grabbed and held tightly. Providing reliable transactions for SkypeOut will make PayPal the favorite pay service of all frustrated SkypeOut users.

I'm amazed Skype let so much ill will fester for so long over MoneyBookers. Since SkypeOut minutes is their first premium service and primary revenue generator, one would think Skype executives would pay closer attention. When your own forums are full of customers recommending competing services because they hate your payment options, some type of action is overdue. Go to work, PayPal.

Once SkypeOut credits are purchased, they appear on your Skype Start page. Whenever you start Skype, the first screen you see shows your remaining SkypeOut credit balance, as you can see in Figure 7-2.

If you don't have any SkypeOut credits, Skype puts a commercial there, which says, "SkypeOut: Global calling at local rates." The rate comment fits. Most calls cost two cents per minute to the two-dozen most common country-calling destinations, and other rates remain reasonable even by broadband phone-pricing standards.

European customers struggle with various VAT (Value Added Tax) rates and details of business within the European Union. So far, for instance, Skype does not offer a way to buy SkypeOut credits without paying VAT. Lucky for me, Skype doesn't charge VAT for those living outside the European Union.

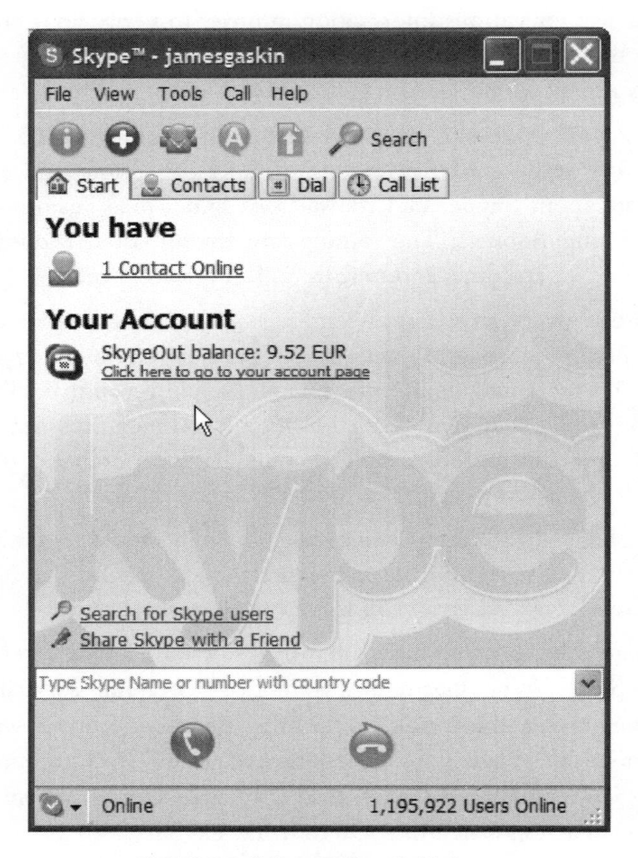

FIGURE 7-2. Notice my SkypeOut balance

Businesses buy things without paying VAT, but Skype's help screens say the service is targeted at consumers who must pay VAT. Presumably the Skype For Business program, when launched, will take care of these details.

Calling Your Non-Skype Friends

Once SkypeOut credits are in your virtual pocket, calling can commence. There are four ways to call using SkypeOut:

- Type the phone number in the text field at the bottom of the application.
- Use the dial pad to type the number.
- Highlight a SkypeOut contact saved previously.
- Re-call a call from your Call list.

Option 2, using the dial pad, is a pain (put your cursor over the 0 and left-click for two seconds to get a plus). Skip that one. But you will have to use the dial pad to respond to requests for phone menus (press 1 for sales, 2 for returns, 3 for incredibly boring music on hold, 4 . . .). Still not fun.

You must use the international dialing format with the "+" and country code before each SkypeOut number called. Americans dial the least amount of digits–for example, +1212-555-0123. After you enter the number in the text field, click the green button with the off-hook icon and SkypeOut will start dialing.

After you've called someone on SkypeOut, you can save them as a Contact. Right-click on that call in the Call List page, provide a name for that saved call, and you're done.

Quality seems, to me, consistently below the Vonage call quality, and well below the quality for a Skype-to-Skype call. Randomly the call quality drops far below normal, more often than on other broadband phone services. But with rates so cheap, hanging up and calling a second time works.

Calls to traditional telephone numbers stay encrypted until they leave the Skype network. Calls to toll-free numbers still incur SkypeOut charges. No 911 or other emergency call options exist.

Tracking SkypeOut Usage

The first trick to knowing your SkypeOut credit can be found on your Skype application, as seen in Figure 7-2. You must visit the Skype web page and log in to see more SkypeOut information. My call list appears in Figure 7-3.

Reach this screen by logging in to Skype, choosing My Account, and clicking either My Recent Calls under the SkypeOut balance listing or the Call List menu item on the left side of the screen. You may also click All My Purchases to track expenditures on Skype equipment such as headsets.

You can't make SkypeOut calls if you don't have enough SkypeOut credit for at least one minute of calling. Skype also stops you from over-buying SkypeOut credits by blocking more purchases until your credit drops toward empty.

If you quit using Skype for 180 days or more, any SkypeOut credit will disappear. Calling again will reset that 180-day counter, but it may take a few days to update your screen.

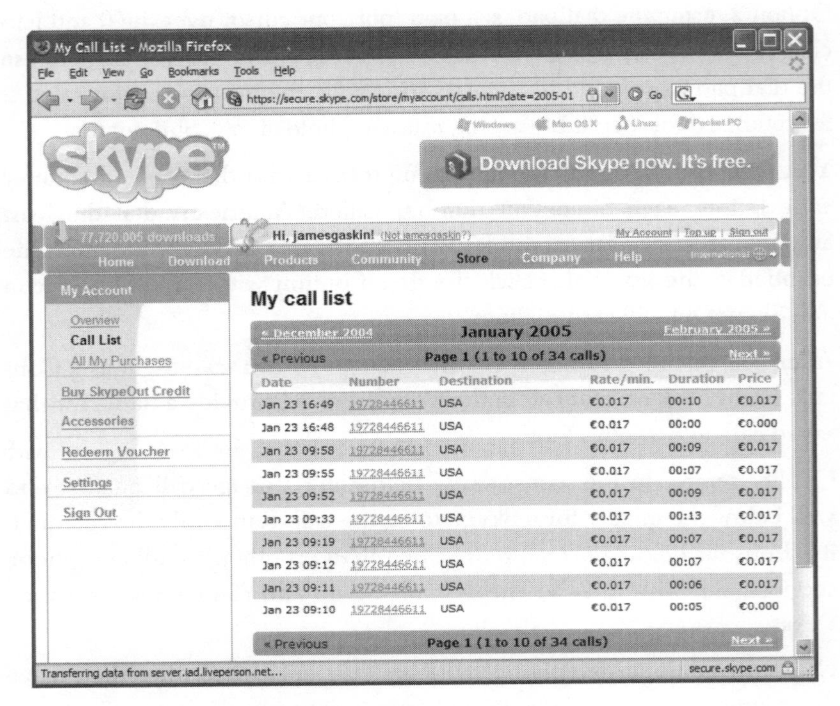

FIGURE 7-3. SkypeOut call detail

Broadband Enhancements to Traditional Telephone Services

Bits are bits, as we've learned. Drilling down a bit more, broadband phone companies can grab the bits that constitute telephone calls, especially the bits at the start of the message detailing the phone number to receive the call and the number (or Internet address) of the caller. Clever people have determined how to watch for calls, grab certain ones, and add value to them.

Phone calls become software, software switches control where data streams (including phone calls) go, and Internet speeds mean a call can be grabbed, rerouted to a special server, and then delivered without degrading the voice quality. A new age of telephone services are developing, and even traditional telephone users benefit.

CallWave

CallWave, Inc., (*www.callwave.com*) precedes the wave of new telephone applications. They have two interesting products, both of which use Internet Telephony in a new way.

Their first product, Internet Answering Machine, uses Internet Telephony to watch your line and grab any call that comes in when your computer is online. They grab the Caller ID information and put that onscreen so you can see whether you want to take that call.

You can listen to the caller leaving a voicemail message and click a button on the onscreen application to talk to that person using a softphone installed as part of the CallWave application. If you prefer, you can redirect that call to your cell phone.

Eight-hundred thousand users feel the $3.95 per month for CallWave offers enough benefits that they don't need a second phone line. Four dollars versus a minimum of $20 for a second line? That does make sense.

Their second product, which they call a pair of products, works with your cell phone. They call them Mobile Call Screening and Mobile Call Transfer.

Mobile Call Screening adds a similar type of call handling to cell phones that CallWave adds to home computers. Using your cell phone, you can see the Caller ID information and decide whether you wish to take the call. If so, press a button and talk. If not, let the caller leave a voicemail message.

If the voicemail sounds interesting, you can pick up that call. Yes, this works just the way your home answering machine works. Screen your calls and pick them up if you want. Cell phones have voicemail, but they don't let you listen to the call and pick it up.

The second part, Mobile Call Transfer, addresses the expense (high) and quality (low) of most cell phone conversations. Ever been sitting right beside your home telephone and get a call on your cell? People tend to call the number most convenient for them, not for you. So you sit and talk on your cell for umpteen dollars while your home phone sits there mocking you.

Using Mobile Call Transfer, you take the next step following the cell call screening. Get a call you want on your cell while you're near your home phone? Hit one button and have it transferred, immediately, to your home phone. The caller never knows you're listening to a better connection for less money than possible on the cell phone they called.

Price? Three dollars and ninety-five cents per month per line. How many times have you gone over your cell phone minutes and paid a premium? Not any more.

CallWave keeps their data servers in a hosting facility in southern Nevada. As mentioned, Internet Telephony works so quickly that a call from north Cleveland can be grabbed, routed to Nevada, and sent back to south Cleveland without any noticeable delay or voice problem. Meanwhile, the receiver in south Cleveland transferred the call from her cell to her landline to save money, and the guy begging for another date has no idea (about the call, but probably not about the date, either).

The CallWave people are nice, and their technical expertise should be put to work on behalf of the 911 issue if the broadband phone companies can't get their act together. Technology pioneered by CallWave and similar advances will drastically reshape the world of voice communications over the next few years.

Teleo

A company and product still in beta testing (although they're taking money, so perhaps beta is a misnomer) may be the SIP-based answer to Skype for Business. Teleo (*www.teleo.com*) provides a softphone that integrates into Microsoft Office applications, making calls an easy one-click option. Millions and millions of people (about 600 million according to Microsoft) run Microsoft Office, including Outlook and Outlook Express. Teleo uses Microsoft programming interfaces to put menu items labeled "Call using Teleo" inside many Microsoft applications.

Bits are bits, right? Integrating with Microsoft applications makes Teleo more business-friendly than any other software phone available before. Get an email from a coworker and need to tell him how wrong he is? Click the name in the From: field in the email, and Teleo looks up the connected phone number from the Outlook database. Ring, refute, hang up.

Many mobile workers live with their laptop, but even they must put down the computer sometimes. During those times, tell the Teleo service to forward all calls to your laptop number to your cell phone.

Yes, your laptop has a phone number, just as it does with the Vonage softphone and the other SIP-based providers such as SIPphone and FreeWorld Dialup. Plugged into your company network in the office?

The Teleo service finds your laptop. Plugged into the wireless network waiting at the airport? The Teleo service finds your laptop. Sneak away to the coffee shop to read your favorite comics (*www.comics.com*) so the people in the surrounding cubicles won't hear you laugh? The Teleo service finds your laptop.

More business friendly, Teleo allows companies to private-label their product. Since many companies have thousands of mobile or remote workers relying heavily on their laptops, that's an attractive offer. The end-to-end encryption provided by Teleo, similar to the encryption with Skype, makes an even stronger sales pitch for some industries and paranoid managers.

Told you bits are bits, and Internet Telephony will turn voice streams into data managed just like any other stream. So pay the $2.95 or $4.95 per month depending on minutes for Teleo and have your cell phone and laptop nag you wherever you are.

Is this really progress? Yes, if you remember to turn off Teleo and your cell phone and let people talk to voicemail when you're laughing in the coffee shop.

Redial

2005 will be the year Internet Telephony starts switching not only consumers, but emergency services and huge corporation's mindsets about voice communications. The ability to energize a phone call the way the computer energized the bookkeeper's journal by creating the spreadsheet will impact every phone user from Aunt Martha to Microsoft to the U.S. Government to Fortune 500 companies. In fact, every Fortune 500 company uses some type of broadband phones already, so the switch is on at all levels.

New ways to talk, call for help, and conduct business zoom toward us. (Or are we zooming toward them? Doesn't matter). Phone flexibility, a feature never before seen, will become commonplace. I just hope we have something worthwhile to say over our new voice conduits with infinite options.

8 TRICKS, TIPS, AND TECHNIQUES FOR ADVANCED USERS

Dig deeper, drill down, discover new ways to make your life easier. That's what advanced users do everyday. I will help you do a few things and hope to point you in a direction that will take you to even more interesting discoveries. And when you do, please send a note so I can share it with the other readers at *www.gaskin.com/talk/*.

Hang on, and you will find some interesting ideas. Some may even solve problems you didn't know you had. Believe me, solved problems are the best ones to find.

If Internet Telephony stirs your blood, and this book merely serves to tease you with the amazing possibilities exploding in the next two years, this chapter will make your yearning worse. Read about some techniques for seriously advanced users. You can even start your own telephone network.

Adding Phone Extensions by Rewiring or Other Options

When I tell people their existing telephone wiring will not work with their new broadband phone, they always react the same way: despair. People expect phone extensions to work because they always have. When they purchase broadband phones, they just assume the new phones will act exactly like their old phones.

Oops, that fine print got us again. The good news? You have options. Some are free, some cost a little, some cost a lot, but you still have options.

The bad news? Most of these options costs money, and one reason many people switch to broadband phones is to save money.

Remember that many broadband phone users keep their traditional telephone and use their new, advanced phone for long distance only and as a standard second line. In such a case, existing extensions can continue to function as they are. The new broadband phone options don't have to be used in any certain way.

This option is the least expensive and sidesteps many of the problems encountered when making your broadband phone service your only phone service. Yes, you can rely completely on your broadband service, but new and growing markets have bumpy roads until market size smoothes the way.

Let me outline three options for handling extensions when switching to broadband phones. These rank in the order I feel provides the best value and highest satisfaction rating for most users. You're free to disagree, of course, since this is your book and your phone. But don't automatically assume your new broadband phones must be handled they way you handled your traditional telephone lines. If you stop using Ma Bell's phones, don't keep letting Ma Bell tell you how to use your phones.

Plan A—Expandable Cordless Phones

Replacing old extension phones with expandable cordless phones works quickly and provides an excellent platform for all your new broadband phone services (refer back to Chapter 4 to refresh your memory and see some pictures). I believe this option provides the highest satisfaction rating with the best new hardware to enhance your new broadband phone service.

If you haven't bought a new cordless phone recently, you may be amazed at their capabilities today. I know I was.

As an example of the modern cordless phone, Motorola sent me one of their MD761 base stations and a MD71 handset to match. The Motorola MD761 base station includes room for four AAA batteries so the phone continues working during home power outages–something else that surprised me.

These phones fairly well represent the features sets of expandable cordless phones today. This unit has an answering machine, but most models have a choice of answering machine or not. Table 8-1 shows the features of the MD761 and MD71.

TABLE 8-1. Looking at the MD761 and MD71

MD761 base station	MD71 handset
5.8GHz FHSS digital technology	5.8GHz FHSS digital technology
Caller message screening	Handset speakerphone
Large lighted message alert	Custom ring tones based on phone book
Handset speakerphone	Fifty private phone book memory locations
Twenty-four selectable polyphonic musical ring tones	Four-line theater style fading backlit display
Custom ring tones based on phone book	Lighted handset keypad

TABLE 8-1. Looking at the MD761 and MD71 (continued)

MD761 base station	MD71 handset
Fifty private and 48 shared phone book memory locations	Phone company voicemail indicator
Caller ID with visual call waiting	Audible and visual low battery indicator
Four-line theater style fading backlit display	Handset to handset intercom and room monitor
Lighted handset keypad	Auto-answer hands-free intercom
Phone company voicemail indicator	Redial, flash, hold, and mute buttons
Audible and visual low battery indicator	Last five number redial
Handset to handset intercom and room monitor	English, French, Spanish, and Portuguese displays
Auto-answer hands-free intercom	Auto channel search before and during call
Redial, flash, hold, and mute buttons	Belt clip included
Last five number redial	
Handset locator	
English, French, Spanish, and Portuguese displays	
Three-way conferencing	
Call progress timer	
Headset capable (2.5-mm plug models)	
Battery backup—4 AAA batteries	
Auto channel search before and during call	
Wall mountable	
Belt clip included	

The FHSS (Frequency Hopping Spread Spectrum) digital technology referenced at the top of Table 8-1 may seem like marketing blather, but it's real technology used by multiple vendors. 5.8 GHz is the highest frequency available for consumer telephone electronics today, up from the original 900MHz cordless phones and the previous 2.4GHz high level. This frequency range won't interfere with, or be affected by, your microwave oven or your wireless network based on 802.11b and 802.11g technology.

In addition, the answering machine component includes the following features:

- Digital telephone answering machine (no tapes)
- Up to 15 minutes or 59 messages recording capacity

- Access the answering machine from handsets and remote telephones (call from other locations)
- Message forwarding
- Answering machine toll saver (adjusts number of rings based on whether messages have been left)

The base station takes more room than those without an answering machine, but not so much that it takes over your desk or end table. Figure 8-1 shows the image straight from the Motorola web site.

FIGURE 8-1. One of the new breed of expandable cordless phones

Expandable handsets obviously take less space. The charging cradle for those is only slightly bigger than the handset itself and takes up little room.

The large round dial-looking thing isn't a dial, but it lights up when the phone rings. It also blinks orange on and off when the phone company has a voicemail waiting (I tested it with Vonage but all the phone-centric providers work).

Normally, you don't know when you have a Vonage voicemail at home until you pick up a handset and hear a stuttering dial tone. That signal means voicemail awaits, although I always think the network's having trouble until I remember about voicemail. The blinking light on the Motorola unit offers a great middle ground between the handiness of getting voicemails from your provider via email for easy remote checks and still announcing voicemail when you're home and want to know there's a message waiting.

Oddly, the orange light doesn't blink when you have a message on the internal answering machine. For that, the number of waiting messages blinks on the display (the top right of the unit) and the expandable handset show the number of messages on their screens. You can hear your messages from any handset or by pushing a button on the base station.

There are two things I'm not thrilled about with this expandable cordless phone. Okay, three. I wish waiting messages in the internal answering had a better alert, more like the waiting voicemails at the phone company. I wish the handsets stood up on their own when not in the cradle, because most other cordless phones do. And I think the wall mounting bracket arrangement is pretty stupid. But those are pretty minor complaints.

I believe you will spend less money overall and get more use from new expandable cordless phones than any other option. But you can feel free to disagree and move on to Plan B.

Plan B—Wireless Phone Jacks

If you have a set of phones you're happy with, or are the type of person who leaves things laying around and can't find them again and so don't trust yourself with cordless phones, Plan B may help. Wireless phone jacks convert voice signals to run over the electrical wiring in your home and allow you to put a phone plug anywhere you have a power outlet.

This type of networking is perfectly safe, and works with fast Ethernet networks between computers as well. In fact, I connect my children's PCs upstairs over the AC wiring because it's just a little too far from the router to work reliably over a wireless connection. And voice phone calls require much less bandwidth than computer networking.

In fact, one of the companies making these products, Phonex, has an excellent line of network adapters that bridge Ethernet across the power lines in your home. See their wireless phone jack products in Figure 8-2.

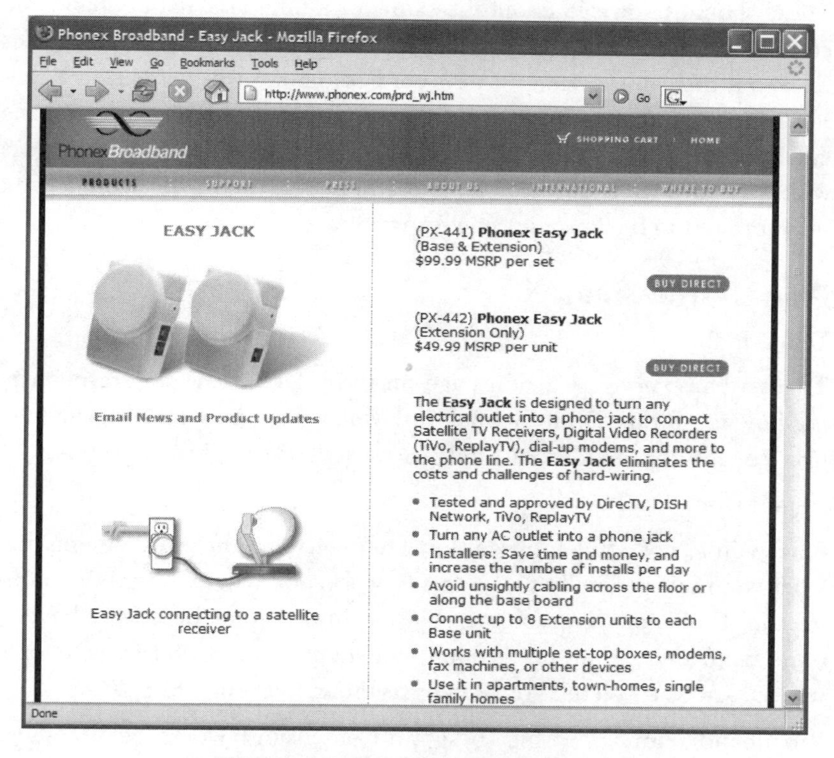

FIGURE 8-2. Easy Jack products from Phonex

The Phonex company (*www.phonex.com*) advertises these units for connecting a satellite dish receiver to a phone line. When you go looking for these units, you probably won't find them in the telephone or computer areas, so look in home networking or satellite sections. They work great for normal phone operations too, of course.

A good explanation of these devices and others that utilize power lines for networking appears at *www.powerlinephones.com.* You may be surprised what you can connect through your power sockets.

I've found these wireless phone jacks listed online at RadioShack and Best Buy:

- *www.radioshack.com*
- *www.bestbuy.com*

Check these locations and anywhere else you feel comfortable buying. Go to any online electronics supplier and search for "wireless phone jacks" and see what appears.

These should be priced about the same, or a little less, than expandable cordless phones. To me, buying wireless phone jacks to keep old phones for about the same price as a new cordless phone with extension handsets is false economy. But it's your economy, so make your own choice and don't look back. There's a good chance expandable cordless phones will drop in price a bit more over the next year or so, anyway, so you may turn out to be the smart one for waiting.

Plan C—Rewiring

Don't do this.

That's what every instruction page on the Internet says before describing how to do this. Even the major players who supply this information, like Vonage and AT&T CallVantage, warn you strongly against this.

Again: Don't Do This.

Since you're not going to listen, I will tell you what the requirements are, what you must generally do for this to work, and where to find the information. I will not give you step-by-step instructions, because I haven't done them myself and don't think you should, either. But some people want to know, so I'll tell you. Not everything, but some things.

Broadband phone rewiring to support extensions works best in these situations:

- Single family dwellings
- A place in which you have complete control over, and access to, the wiring
- A place in which there are no alarms or medical alert equipment connected to the phone line
- A place in which there is no satellite TV, TiVo, or other Digital Video Recorder connected to the phone line

You can work beyond these situations, but the process includes a much higher degree of difficulty. If you're in a multifamily dwelling, for instance, finding the demarcation point between the traditional telephone company and your unit may be impossible. If you have to break the law–i.e., trespass to reach the wiring connections–you can't do this.

One reason you shouldn't do this in a multifamily dwelling is because you don't have control over the wiring. If someone later comes and reconnects the traditional telephone outside wiring to your unit, the voltage from the phone company may fry (yes, cause devices to break down or even burst into flames) your equipment. Okay, frying and bursting into dramatic flames is unlikely, but possible. Things will probably just quit working, but you never know.

Those with alarm systems connected to their phone lines have extra hurdles to jump. Chapter 7 discussed these issues, so go back if this applies to you.

If you have medical equipment connected to your phone, or medical equipment that needs occasional access to a traditional telephone line, don't risk it. Add a broadband phone line for all the good reasons listed previously, but keep the traditional telephone line for safety's sake. I shouldn't have to mention this, but sometime, somewhere, somebody with a heart monitor hooked to his poor, sick mother will make news by adding a broadband phone and mishandling the 911 activation instructions. Will it be blamed on the idiot who disconnected his mother's lifesaving heart monitor from the telephone? No, it will be blamed on the broadband phone company.

On the other hand, if you mess up your TiVo or satellite TV Pay Per View connection, I couldn't care less. People watch too much TV anyway. Go to a concert.

Broad steps

There are five broad steps for this project:

1. Disconnect your inside wiring from your connection to the traditional telephone company.
2. Verify that your internal phone lines have no voltage load on them.
3. Plug any phone connection into the telephone port on your broadband router or telephone adapter.
4. Plug in the rest of your extensions.
5. Use your old phones just like normal on your new broadband phone service.

The reason any phone plug will work for your broadband router or telephone adapter connection? Your home telephone wiring runs in parallel, meaning all connections link to all other connections. Essentially, your home phone wiring is one really long wire pair running all through the house.

Retrofit Requirement

Make sure you can easily reconnect the traditional outside telephone line when you sell you home.

Because the lines are in parallel and a limited amount of voltage powers all the phones on the circuit, you will cause problems if you connect too many phones to the line. Most recommendations say a traditional telephone line can handle four or five physical phones before the load drops the voltage too low to kick off the ringer when a call comes in. Some folks report adding more phones with no problem, but your mileage may vary. Short runs with high quality cable and connectors will power more phones than long runs with poor cable and interference from electrical wires running close by.

Feel comfortable with these broad steps? Then go forward, but you go without my blessing.

Detailed instructions

If you understand these steps but don't want to do this yourself, show this information to an electrician or telephone service person. You will spend money but save frustration, and the job will likely be done with higher quality.

1. Find your demarcation point where the telephone line connects to your home. This should be a gray, green, or black plastic box on the side of your house or in the garage or basement. This box is called a Network Interface Device or Network Interface Unit.
2. Open the consumer half of the Network Interface Device and disconnect the telephone line. There are multiple ways to do this depending on the type of connector, the number of wires, and the phases of the moon. Be careful!
3. Indicate the telephone lines have been removed on purpose so some well-meaning technician doesn't plug them in again. Roving bands of good deed technicians are rare, but just when you think you're safe, they charge over the horizon waving new wires, connectors, and screwdrivers.

4. Verify the house wiring no longer connects to the telephone company and nothing on the line, like telephone signal boosters, is still providing voltage.
5. Plug a phone cord between a phone connector in the house and the phone plug on the broadband router or telephone adapter.
6. Check to see if all your extensions give a dial tone when picked up.
7. Check to see that all your extensions ring when called.
8. Cross your fingers and hope no one plugs the traditional telephone line back in when you're not looking.

Working with homes wired with more than one line complicates the instructions. Homes too old to have a Network Interface Device will need to be retrofitted by your traditional telephone company before rewiring. Don't tell them what you're doing because it will hurt their feelings.

Wiring in large homes may be split depending on the situation and your builder. If it is, your wiring hassle will double because of differences at the Network Interface Device and at every phone jack. You will need line splitters (effectively a two-plug molded plastic piece that separates line one from line two) for each extension. Enjoy.

Since you haven't learned your lesson, here are some resources:

Vonage Knowledge Base

www.vonage.com/help_knowledgeBase_article.php?article=649

AT&T CallVantage Home Wiring Do-It-Yourself Guide for single line phone installations

www.usa.att.com/callvantage/assets/pdf/acs_hwg_single_line.pdf

AT&T CallVantage Home Wiring Do-It-Yourself Guide for phone installations with two or more lines

www.usa.att.com/callvantage/assets/pdf/acs_hwg_two_lines.pdf

One of the better help sites with good tool explanations

www.jakeludington.com/ask_jake/20050206_voip_over_phone_wiring.html

As do many decisions in life, your method for handling extensions comes down to the balance between cost and convenience. Expandable cordless phones are more convenient and cost more. Rewiring, if possible in your situation, costs less but is also much less convenient.

Personally, since you're getting a new phone service, I think you should get new phones, too. That's because I think you're special and deserve the best.

Improving Your Call Quality

When your traditional telephone line sounds bad, you have little recourse. Okay, *no* recourse. But you can affect the sound quality of your broadband phone connections. Most of the recommendations from providers focus on you not messing up the voice quality, because they believe their quality sparkles.

Manage Your Bandwidth

For high quality conversations over a broadband phone of any type, you need about 90 Kbps. When you're listening, that slice of bandwidth comes from the downstream portion of your connection, and when you're speaking, you need room on the upstream portion of your connection.

Many technical people believe "broadband" means at least 2 Mbps (or 2,000 Kbps) of bandwidth. Some even believe you shouldn't call it broadband until it's at least 10 Mbps (10,000 Kbps). European and Asian countries agree with those high speeds, but the U.S. Federal Communications Commission allows vendors to call DSL "broadband" at 384 Kbps downstream and 128 Kbps upstream bandwidth. And these companies do that with a straight face.

Getting More Than You Give

For a variety of technical reasons (and to reduce their cost of equipment), broadband service providers deliver different speeds downstream (to you) and upstream (from you). Ratings appear, in small print, like 384 Kbps/128 Kbps or 3 Mbps/1 Mbps. The larger (first) number is your downstream bandwidth while the smaller (second) number is your upstream bandwidth. Few residences have a symmetrical broadband connection (the same speeds upstream and downstream) although that is common for businesses.

If your broadband connection upstream is above 512 Kbps, you can have two or three broadband phone conversations online at once with no problem. If your upstream connection is in the 1Mbps range, as is the case with many cable providers, you're in great shape for almost any residential broadband phone application, including multiple calls concurrently.

If your broadband connection upstream is limited to 128 Kbps, you can have a single broadband phone conversation with excellent quality. But if someone else in the house starts anything that demands a lot of bandwidth from your Internet connection, such as playing an online game or uploading files, your call quality will drop.

Easy answer? Upgrade your broadband connection. Unfortunately, that's not always possible. Worse, when you upgrade, you spend more money on your connection, somewhat negating the savings you get from switching to broadband phones.

Typical answer? If the bandwidth siphoned away from your broadband phone connection drops your call bandwidth too low, the call may drop. In that case, especially with phone-centric providers, you need to lower the phone's bandwidth demands so the call will stay connected at lower bandwidth.

Vonage offers a tool called Bandwidth Saver to handle exactly this type of situation. You can find this screen when you log in and click on Features (the menu choice just to the right of the middle of the menu choices showing in Figure 8-3) and choosing Bandwidth Saver. They have configured three settings: normal, higher, and highest quality. The default is to the highest quality.

Looking at Figure 8-3, you can see normal sound quality is 30 Kbps, higher is 50 Kbps, and highest is 90 Kbps. If you believe you're getting cheated by using only 30 Kbps for the normal quality phone settings, go back to Chapter 1 and reread how digital telephone conversations save bandwidth and improve call quality.

Choosing a lower setting will allow your available bandwidth to better maintain your call connection while still providing adequate support for other demands. I won't lie to you and tell you that a heavy load on your Internet connection will never degrade your call quality, because it will have an impact. But by adjusting your call's bandwidth requirements, you can keep the call from losing so many packets in the conversation that the connection drops out, and that's an improvement.

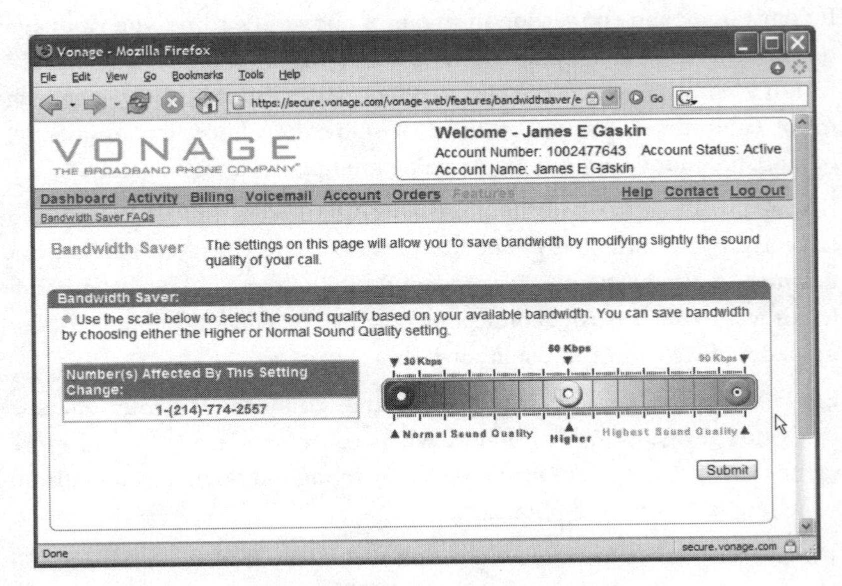

FIGURE 8-3. Adjusting Vonage call quality

Upgrade Your Headset

Skype and other softphone users can make an enormous improvement in call quality by upgrading your headset. And if you don't have a headset, get one.

Inexpensive headsets are engineered to deliver the frequency range needed for traditional telephone line voice calls. The frequency range provided by Skype dwarfs the traditional telephone call range, but cheap headsets won't reflect that improvement.

You won't hear as much improvement with a better headset in your calls from the softphone to traditional telephone lines, because the traditional line drags down the quality. But a good headset on each end of the conversation–meaning a good microphone for someone to speak into and a good speaker to relay the sound–makes a Skype call sound amazingly lifelike.

Quality of Service Technology

Remember the mentions of Quality of Service back in Chapter 4? That technique allows routers to prioritize packets based on their information. By fall 2005, some of the new routers delivered by Vonage and other phone-centric providers will include Quality of Service technology. These traffic-control techniques work only when both ends of the connection work together, so the other phone providers must retrofit their end to match, as well.

Business broadband phone services include this technology. Small business and residential routers sometimes have Quality of Service options, but today those provide value only when two similar routers communicate. But as Quality of Service technology rolls downhill to affordable home routers, voice packet traffic control will increase, as will the quality of sounds for all broadband phone conversations.

Conference Calls

Want to talk to more than one person at a time? You can do so with little extra trouble.

Vonage and the other phone-centric providers all offer three-way conferencing, which allows you to call one person and then call another and link them into the conversation. Of course, many of the traditional telephone providers allowed the same thing.

Four steps get your Vonage three-way calling session started:

1. Call the first person.
2. Once connected, hit your Flash button just like you do for Call Waiting.
3. Call the second person.
4. Once the second person is connected, hit the Flash button again to join the two calls.

If you think this looks amazingly like a clever way to utilize an advanced Call Waiting feature, I have to agree with you. No matter the technology behind this, it can be handy. This works only for calls within the United States and Canada, however, because of how other phone systems handle the Flash button signals.

Skype allows up to five total conference call users (the conference initiator plus four invitees). Only one of the conference members can be a SkypeOut connection. You can use the Tools menu to start a conference call with all involved, or add people to an ongoing call.

One person must be the conference host and invite the others. Because each call adds to the bandwidth requirements for the host, pick the host based on their equipment. A host with a fast computer and high bandwidth connection make a five-way conference call much more pleasurable.

Figure 8-4 shows the screen that appears after choosing Tool → Create Conference in your Skype application.

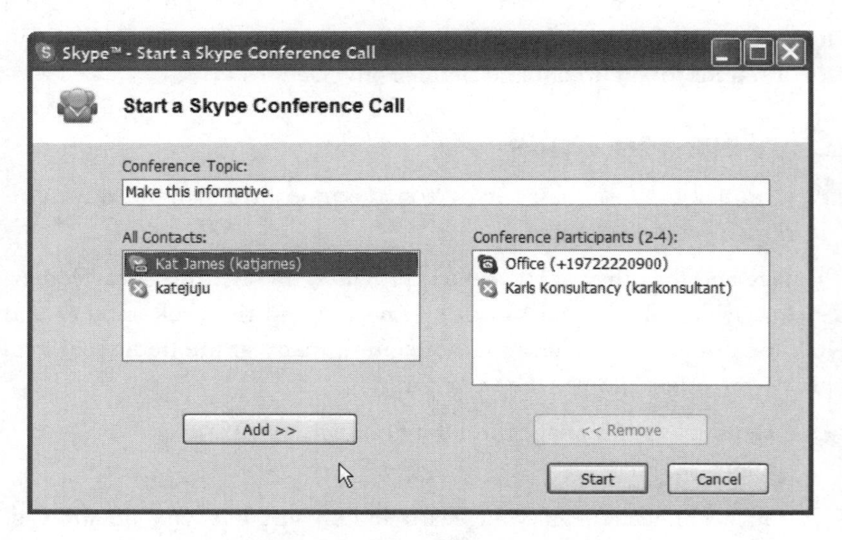

FIGURE 8-4. Adding a third person to the list

Contacts appear in the left window, and participants appear in the right window. Highlight one or more contacts, then click the Add button (the cursor points there) to move them into the participants list. Once you have your participants selected, click the Start button.

When each contact answers, all contact icons appear in your Skype application. The names scroll sideways if they're too long, which is why KarlsKonsultancy doesn't show up completely in Figure 8-5.

Skype shows call duration below the host's icon. Notice there are SkypeOut and regular Skype contacts mixed in this conference call, a nice feature. Sometimes adding the SkypeOut call introduces a bit of echo, but if it's distracting, hang up on that participant and call them again while the others wait.

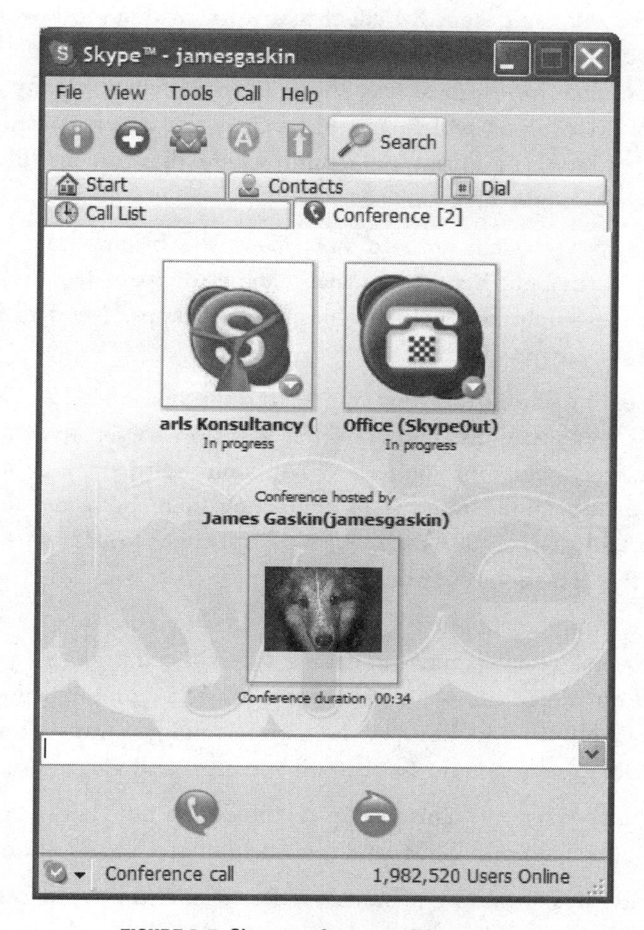

FIGURE 8-5. Skype conference call in progress

Families love this feature. Call Mom on Mother's Day and have three other siblings on the call, all for free if you're all on Skype. Saving money on the call will be important because I'm sure your Mom's present cost plenty.

Skype Instant Messaging Tricks

Instant Messaging as an advanced trick in a broadband phone book? Yes, when it's part of the world's most popular broadband phone network.

There are two important tricks for advanced users available with Skype Chat. One of these tricks may save your some aggravation one day.

First, Skype offers a decent IM client that works with almost every company's network yet avoids the bad reputation of AOL IM, MSN Chat, and Yahoo Chat for business use. If you have a small company and see where an IM product will help you, such as for quick questions and answers between employees no matter where they are in the world, Skype gives you one for free.

Within the Skype Chat application, a complete history stays on your hard disk by default. Better than that, each participant has that history on their own computer, making automatic backups. Need to refer to something? Everyone has it.

You can export the history text and search the file, offering even more value in some situations. Did Alex tell you how to set up your Word styles two weeks ago, but you forgot? Did Laura send you a phone number you need again? Rather than contacting them both and admitting you're lost and or forgot what they told you, check your Skype IM history and then go about your business.

Second, a company named Connectotel (*www.connectotel.com*) tested a text-messaging gateway between certain types of cell phones and Skype. Designed for GSM (Global System for Mobile communication) cell phones, widely used in Europe, the Connectotel gateways allowed cell phone users to send text messages to Skype users and vice versa.

Why do this? Why not? This just opens the door a little between the digital cell phone network (GSM) and the world of broadband phones.

Admittedly, this would be a much smaller deal if the GSM cell phone users could just call the Skype user through the Skype phone network. But until SkypeIn debuts, you can stick with text messaging or switch to Teleo so calls to a person's laptop would work.

Broadband Phone Politics

How serious are some telephone companies taking all these broadband phone companies? Pretty seriously. Add in the fact that the government-run traditional telephone company remains the only company in many repressive countries, and you can have some bizarre laws.

Widely reported in late February by such news organizations as CMP (*www.techweb.com/wire/networking/60403862*), broadband phone users in Costa Rica may face serious consequences. The national telephone company's revenues have dropped because of peer-to-peer telephone connections over the Internet. When reported that 20% of the country's international calls go over broadband instead of the state-run telephone monopoly, bureaucrats got nervous. And nervous bureaucrats in countries with government-owned telephone monopolies tend to make things they don't like illegal.

Whether the "clamp down" camp will win over the "it makes more business possible" is unknown. But even in the U.S., those two arguments rage constantly.

If you're tired of company politics with your telephone company, if you want to get involved in such, start your own broadband phone company. You can do it with softphones and a little programming.

Check out "the on-line reference for Internet telephony" called iptel.org at *www.iptel.org*. Anything you can't find out about Internet Telephony there can probably be found at SIP Foundry (*www.sipfoundry.org*). Software toolkits, source code, and cutting edge research in making all this work await the curious.

Once you get your phone company up and running, you may need a PBX (Private Branch eXchange), just like the big companies have. No sense trying to buy a traditional telephone switch in the new century, so go to Digium Inc., and check out Asterisk (*www.asterisk.org*). Some people, although not what you would really call normal people, run one of these Linux-based phone switches in their home for everyday use (not me, I promise, but I know some of them). And if you have a small business that needs new phone equipment, Asterisk will do everything you need for a fraction of the price of a traditional telephone vendor system. For more information, see *Switching to VoIP* (O'Reilly).

Redial

You have three options for populating your home with broadband phone extensions, and I hope you choose a safe one. You have new ways to improve your call quality–always a good idea. And that's even easier when you have new extension phones with new technology and handy features.

If you get enthused, there are exiting new ways to use and abuse your phones. You can even start your own phone company and be the president, or at least play one on TV.

9
GO WIRELESS

People love feeling unbound, and they express that freedom when they buy telephones. People love cell phones, cordless phones, speaker phones, and tiny Bluetooth headsets that use short range radio waves to connect to the cell phone on their belt.

Businessmen Are Nuts

Does it strike you as contradictory when a man says he loves the freedom of wireless phones while tightening his necktie? His phone should be free but his head tied down?

I focus on wireless connections, unbound communications, and coffee shop Computer-centric conversations over Wi-Fi in this chapter. Some of these wireless phones work at home and some on the road. All provide satisfaction for those who like to walk and talk.

Let me start with cordless phones for home and small business use, then wander farther afield (figuratively and literally). I'll even tell you about the ultimate free long distance call made from 35,000 feet using Wi-Fi in a new airliner through a laptop via Skype.

Skype and SIP-Friendly Cordless Phones

I've already discussed the value and perhaps necessity of cordless phones with phone-centric broadband phone providers like Vonage, but the Computer-centric crowd really needs the help. Millions of cordless phones fit the plug on a broadband phone router, but only recently have some been ready to plug into a computer.

As broadband phones continue to surge in popularity, keep an eye out for new phones. The manufacturers are retro-fitting existing phones quickly, and a new world of USB cordless headsets is on the way with enticingly low prices.

Olympia Cordless Phone with USB Plug

Enter Skype and a deal with Olympia, one of Europe's largest manufacturers. Figure 9-1 shows a photo and description of the new phone.

FIGURE 9-1. Works with Skype and traditional telephone lines

You can reach this screen from the Skype store web page, or go directly to *www.dualphone.net*. Shipping charges to locations outside the European Union may be high, because the first currency listed is the euro.

This really provides a great solution to the "tied to the computer" complaint about Skype. It also addresses the pain of multiple phones scattered around your desk. This phone plugs into both your computer with a USB connector and into the traditional telephone line plug you've always used.

There should be no confusion during use to cause a problem. When the phone rings, answer it. The call may be coming in over Skype or via your old-fashioned phone. But it all rings on the same phone.

When you dial out, your Skype contacts show up in the display, including the correct icon showing when they are online. To call a Skype contact, highlight that entry and press the call button. This feature means you don't have to go to the computer even to dial a contact–which you must do if you use an adapter to connect a normal cordless phone. You are truly untethered with this phone.

To call using SkypeOut, dial the number just as you would from the computer. To dial using the traditional telephone line (for example, for local calls to avoid the SkypeOut fee), press the traditional telephone line button and dial.

One charger, two cords, no confusion. This doesn't yet work for Macintosh, but it does include Skype software and the needed utilities to communicate between Skype's application on your PC and the telephone.

Want to double your fun? This unit is expandable, and you can purchase up to four more handsets (*www.telestore.dk/shop/default.asp*). You can make a Skype call through one handset while someone else makes a traditional telephone call on another handset. That's flexible, and wireless.

The only other wireless Skype phone that comes close to the Olympia is a wireless USB headset from Plantronics. You can't dial out except at the computer, but this unit does accept calls from Skype or any other softphone. Figure 9-2 shows a good close-up.

Whether this cordless headset makes more sense than buying an adapter for an existing cordless headset is your decision. I can vouch that Plantronics makes just about the best headsets, although I haven't tested this one. Go to *www.ahernstore.com* for this unit and many other specialized phone products.

Adapters for Computer-Centric Softphones

Just as the phone-centric providers use a telephone adapter, so can the computer-centric providers. Linking standard telephones–or in this chapter, cordless phones–through an adapter to a computer via the USB port no longer classifies as a struggle.

There are Skype-specific adapters, as you can see in Figure 9-3, from a specialist phone product company at *www.rapidvoip.com.*

There are more generic USB to traditional telephone adapters, and the entire product line from all vendors gets smarter with each new version. On their web site (*sipphone.com/callinone*), SIPphone offers a Call-In-One adapter that handles SIP and traditional telephone calls (dial normally for traditional telephone calls and put a # before SIP numbers).

A company called Sysgration introduced their SkyGenie in early 2005–it's an adapter that does exactly the same thing for Skype calls, except you put in ## (two number signs) before the Skype phone number

FIGURE 9-2. Tiny headset for softphone flexibility

(*www.sysgration.com*). A company named ActionTec (*www.actiontec.com*) has a nearly identical product. No doubt many of the huge telephone product companies are waiting for the small companies to prove the market before jumping in (if history continues to foretell the future).

Wi-Fi Internet Phones

One day, someone asked an interesting question: why do I have to leave my broadband phone calling advantages at home? Why can't I take them with me to the coffee shop?

Now you can. Several vendor phones let you dial to your broadband phone account through any public Wi-Fi network. When you reach a hotspot, the Wi-Fi phone connects to the wireless network and reaches your phone provider to initiate your calls.

FIGURE 9-3. Skype USB adapter with extra features

Look for broadband phone service providers to private-label these and "customize" the phone to their network. While that means easier setup, it also almost certainly means the phone will be locked so you can only use it with that one broadband service provider. Yes, that's what your cell phone service provider did to your cell phone, too.

First-generation Wi-Fi phones took quite a while to configure and get working through one of the few providers available. If these phones are going to attract a huge audience, they must be easier to configure and start using. This may be the impetus to lock Wi-Fi phones to providers, just to sweeten that "Out of the Box" experience and help people start using their phone quickly.

Dual-mode cell/broadband phones are on their way as well, letting you choose between a broadband phone service and your cell phone service. The idea of going over your cell phone minutes may disappear if the SmartPhones (as they are called by some) automatically grab a public Wi-Fi connection before grabbing the more expensive cell phone network.

Cell phone networks have the coverage advantage today, but keep an eye on a technology called WiMAX. This technology supports high-speed wireless data networking up to 31 miles in diameter from one tower. The prestandard WiMAX vendor TowerStream (*www.towerstream.com*) has done the best job of rolling out coverage. I've seen their installation in New York City, and they have networks in Providence, RI, Chicago, and Boston, and more plans than I can cover here. Very cool stuff.

ZyXEL Prestige 2000W Version 2

The first Wi-Fi phone I saw came from ZyXEL and was their Prestige 2000W. This unit has been relabeled by other providers, and they sometimes call it a WiSIP (Wireless SIP) phone instead of Wi-Fi phone. You will also hear the term VoFi for VoIP over Wi-Fi (and I didn't think the acronym lovers could make a term uglier than VoIP. Live and learn).

ZyXEL's first version looked like a small cordless phone, but their new version aims to emulate a cell phone. Figure 9-4 shows the ZyXEL 2000W Version 2 in a pre-release photo.

FIGURE 9-4. Ugly to see VoIP on the label, but could be interesting

ZyXEL hasn't made huge inroads, especially when speaking about phone brand name awareness (they make some nice home and small business networking products as well). But they have opened the market enough to gather some competition.

UTStarcom F-1000

Vonage signed a new deal with UTStarcom and their still unreleased F1000 Wi-Fi handset in January 2005. Another cell phone–modeled device, the F1000, will come, no doubt, locked to Vonage. I feel comfortable saying that because Vonage forces Linksys to lock their routers, I can't imagine Vonage being nicer to UTStarcom.

This phone and the ZyXEL look like the "candy bar"–style phones popularized by Nokia. With the market moving toward clamshell designs (one of the reasons Nokia has lost market share is because people like to put phones in their pockets and clamshells protect the buttons and screens better), I hope the Wi-Fi phone manufacturers can shrink their designs fast enough to ratchet up the cool factor and get some media attention.

Calling Considerations

Remember all the discussions about 911 calling? How in the world are providers supposed to handle this now if you have a Wi-Fi phone? Skype and the other computer-centric providers have zero experience with 911 service connections. Vonage and Packet8 lead the phone-centric pack, but neither has service rolled out everywhere.

Cell phone providers have only partially solved the 911 problem for mobile phones, and they own their complete network and have control over all the components (even when they combine service with other cell phone providers in some areas). How is a Wi-Fi phone service provider going to be anywhere nearly as good as the cell phone companies?

Do not confuse your Wi-Fi phone with a cell phone for 911 purposes. It will be years before the cell phone providers get the 911 coverage they're supposed to have, but years after that will pass before Wi-Fi phones have a shot of any decent 911 coverage.

On a more relaxed subject, calling and being called on a Wi-Fi phone should be just as convenient to you (and annoying to those around you in public) as using a cell phone. Since Skype hasn't introduced their SkypeIn service yet, the SIP phone service providers may be able to drive some sales into that hole in Skype's market dominance.

Skype has announced a deal with Motorola to preload Skype software on combination Wi-Fi cellular phones. This agreement also extends to Bluetooth headsets and speakerphones. No specific products, either new or updated, have been released. The initial press release from Motorola promised products in the first half of 2005. We'll see if this book or their products see daylight first.

When they get their Wi-Fi phones out, Skype will have a ready-made marketing advantage, at least in the U.K. They've made a deal with wireless broadband provider Broadreach to allow free Wi-Fi phone (and PDA and laptop) calls from the 350 U.K. hotspots. Normally Broadreach charges over $5 per hour at their hotspots, but not for Skype use. And you thought I was joking about taking your laptop to the coffee shop and calling your friends.

Futures

Imagine how our surrounding air will fill with even more electromagnetic products and service in the next two years. Already we're bathed in radio and TV signals when we leave the house. Cell phone frequencies just about cover the entire country. Wireless network protocols now compete for space in that same soup.

The idea of "leaving" your connections will disappear. What will you leave? Your PDA or laptop will wirelessly connect to any hotspot. Your cell phone awaits any call, no matter how tense the play or quiet the music at the concert.

Now magnify these connections by your wireless network and therefore all the presence improvements added by Skype and their competitors. Why put a camera in a cell phone when you can now put Wi-Fi in a camera? Why not hardcode your home server or email address and mail your snapped images to yourself immediately from any public hotspot?

The first report of a Skype phone call from high over the Atlantic in a commercial airliner appeared throughout the Skype community on February 20, 2005. Boeing's new Connexion service in an SAS airplane provided the Wi-Fi connection, but just enough for one or maybe two callers at a time because of the limited uplink bandwidth.

Luckily there was only a single caller, using Skype, speaking into the microphone of his Mac Powerbook. In other words, he used the worst case–calling scenario for talking privately, although he didn't have to share the limited bandwidth. Yet the report claimed clear Skype quality, making it better than one of those seatback cell phones that cost so darn much per minute.

Where next? Skype already reported calls from the top of Mt. Everest (although I bet they mean the last camp before the summit). Laptops and PDAs make calls, as well as Wi-Fi phones, either dedicated Wi-Fi or Wi-Fi/cell phone combinations.

When you're curious about whether something happened or soon will, keep an eye on the SummitCircle web site shown in Figure 9-5. They seem to keep up as well as anyone.

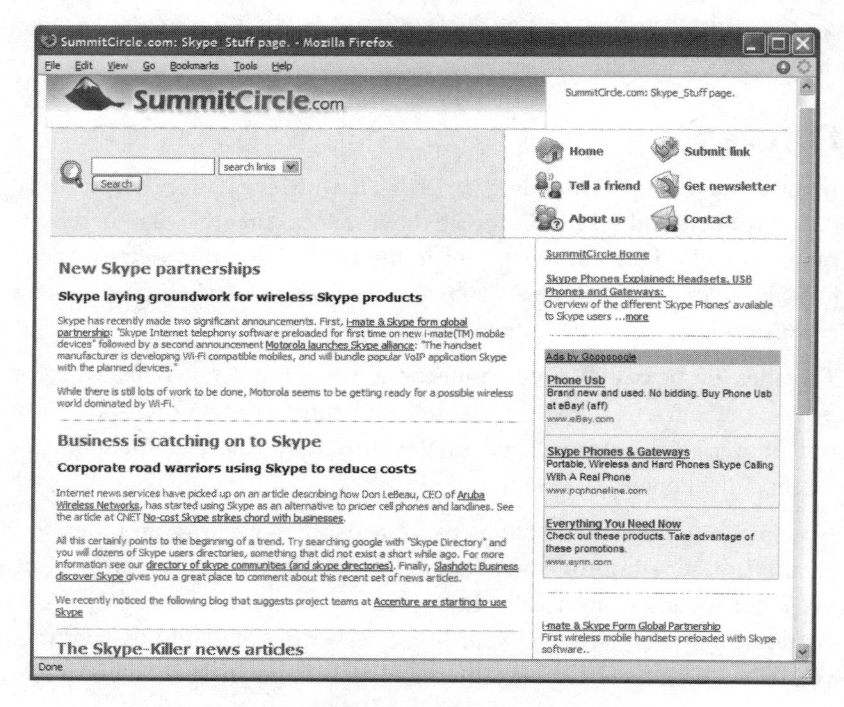

FIGURE 9-5. Third-party Skype news updated daily

One last rumor, no confirmation possible, before we close: Google. They placed an ad asking for someone with experience in Internet Telephony, and the broadband phone community went crazy. Google refuses to comment, just as they always refuse to comment until they're ready to announce a new product.

If it happens, it will do two things: push computer-centric Internet Telephony into the mainstream news cycle, and force companies doing business online to have "operators standing by" on their Computer-centric network for all the users who click their number on a Google ad in some web site. Both of these things are good, but unsubstantiated.

Who knows the future? I know Internet telephony will go more mainstream and get cheaper still, and vendors will provide the products necessary to improve the user experience. I know more value-added services, such as the CallWave discussed in Chapter 7, will be coming soon.

I also know that talking about it helps change people's minds and makes these steps closer to reality. And talking about the future of Internet Telephony over a broadband phone connection pulls the future that much closer.

You will switch one of these days. Go ahead and switch to a broadband phone today, and you will start saving money and get better control over your telephone than ever before.

Redial

New phones advance, old service barriers drop, and broadband phones fit into your pocket and compete with your cell phone. Everywhere in the world, and at the top of it, links to everywhere else through the beauty of Internet Telephony.

Telephones, or at least some type of voice service link, will be in just about every intelligent device in a year or two. Let's just hope our conversations become just as intelligent as the equipment carrying them across the world for free or darn cheap.

Index

We'd like to hear your suggestions for improving our indexes. Send email to *index@oreilly.com.*

T

U

V

W

X

Z

About the Author

James E. Gaskin has been solving computer and network problems for businesses small and large since 1984. He also writes books, articles, and jokes about technology *and* real life. In 16 books and hundreds of articles, network consultant Gaskin tells people faster, cheaper, newer, and smarter ways to connect to each other and the world.

A founding member of the Network World Lab Alliance, Gaskin focuses on small office issues and product testing. He is the leading voice for the small business market through his long-running Small Business Technology newsletter distributed by Network World. As a speaker, Gaskin travels the country in Hawaiian shirts helping people laugh at their IT pain while teaching them new ways to put technology to work. In venues such as Network World Technology Tours and seminars at leading industry conventions such as Networld + InterOp, Gaskin delivers objective information stripped of marketing hype in his trademark "unbiased and unboring" style.

When avoiding computers, Gaskin plays classical chamber music on the violin and tennis (but not at the same time) in the Dallas area.

Colophon

Our look is the result of reader comments, our own experimentation, and feedback from distribution channels. Distinctive covers complement our distinctive approach to technical topics, breathing personality and life into potentially dry subjects.

Mary Brady was the production editor and proofreader for *Talk Is Cheap*. Nancy Reinhardt was the copyeditor. Matt Hutchinson and Claire Cloutier provided quality control. Julie Hawks wrote the index.

Scott Idleman/Blink designed the cover of this book. Karen Montgomery produced the cover layout with Process Type Foundry's Stratum. Marcia Friedman designed the interior layout. This book was converted by Andrew Savikas to FrameMaker 5.5.6 with a format conversion tool created by Erik Ray, Jason McIntosh, Neil Walls, and Mike Sierra that uses Perl and XML technologies. The text font is Berthold Baskerville; the heading font is Adobe Helvetica Neue; and the code font is LucasFont's TheSans Mono Condensed. The illustrations that appear in the book were produced by Robert Romano and Jessamyn Read using Macromedia FreeHand MX and Adobe Photoshop 7.

Keep in touch with O'Reilly

Download examples from our books

To find example files from a book, go to: *www.oreilly.com/catalog* select the book, and follow the "Examples" link.

Register your O'Reilly books

Register your book at *register.oreilly.com*
Why register your books? Once you've registered your O'Reilly books you can:

- Win O'Reilly books, T-shirts or discount coupons in our monthly drawing.
- Get special offers available only to registered O'Reilly customers.
- Get catalogs announcing new books (US and UK only).
- Get email notification of new editions of the O'Reilly books you own.

Join our email lists

Sign up to get topic-specific email announcements of new books and conferences, special offers, and O'Reilly Network technology newsletters at:

elists.oreilly.com

It's easy to customize your free elists subscription so you'll get exactly the O'Reilly news you want.

Get the latest news, tips, and tools

www.oreilly.com

- "Top 100 Sites on the Web"—PC Magazine
- CIO Magazine's Web Business 50 Awards

Our web site contains a library of comprehensive product information (including book excerpts and tables of contents), downloadable software, background articles, interviews with technology leaders, links to relevant sites, book cover art, and more.

Work for O'Reilly

Check out our web site for current employment opportunities:

jobs.oreilly.com

Contact us

O'Reilly Media, Inc.
1005 Gravenstein Hwy North
Sebastopol, CA 95472 USA
Tel: 707-827-7000 or 800-998-9938
(6am to 5pm PST)
Fax: 707-829-0104

Contact us by email

For answers to problems regarding your order or our products: **order@oreilly.com**

To request a copy of our latest catalog: **catalog@oreilly.com**

For book content technical questions or corrections: **booktech@oreilly.com**

For educational, library, government, and corporate sales: **corporate@oreilly.com**

To submit new book proposals to our editors and product managers: **proposals@oreilly.com**

For information about our international distributors or translation queries: **international@oreilly.com**

For information about academic use of O'Reilly books: **adoption@oreilly.com**
or visit: *academic.oreilly.com*

For a list of our distributors outside of North America check out: *international.oreilly.com/distributors.html*

Order a book online

www.oreilly.com/order_new

Our books are available at most retail and online bookstores.
To order direct: 1-800-998-9938 • *order@oreilly.com* • *www.oreilly.com*
Online editions of most O'Reilly titles are available by subscription at *safari.oreilly.com*